Swarm Intelligence and Evolutionary Computation

Theory, Advances and Applications in Machine Learning and Deep Learning

Georgios N. Kouziokas

Lecturer, School of Engineering
University of Thessaly, Greece

CRC Press
Taylor & Francis Group
Boca Raton   London   New York

CRC Press is an imprint of the
Taylor & Francis Group, an **informa** business

A SCIENCE PUBLISHERS BOOK

First edition published 2023
by CRC Press
6000 Broken Sound Parkway NW, Suite 300, Boca Raton, FL 33487-2742

and by CRC Press
4 Park Square, Milton Park, Abingdon, Oxon, OX14 4RN

Library of Congress Cataloging-in-Publication Data (applied for)

ISBN: 978-1-032-16250-8 (hbk)
ISBN: 978-1-032-16251-5 (pbk)
ISBN: 978-1-003-24774-6 (ebk)

DOI: 10.1201/9781003247746

Typeset in Times New Roman
by Radiant Productions

Preface

The application of Artificial Intelligence (AI) has been greatly developed in many scientific sectors the last years, with the development of new AI-based algorithms and optimization techniques. Swarm intelligence and evolutionary computation are generally utilized as advanced methods to solve several kinds of computational optimization problems especially when there is only a small amount of relevant information.

Swarm intelligence was inspired by studying the swarm behavior of animals in nature. The social intelligence of natural swarm behavior was suitably transformed to develop computational optimization algorithms. Swarm optimization algorithms have a stochastic nature where the swarm individuals are traversing a predefined search space while the swarm algorithm evaluates the individual and the global best fitness function values in every step so as to estimate the next positions of the individuals in the space depending also on predefined hyperparameters and conditions.

Evolutionary computational optimization was inspired by the natural biological evolution. Evolutionary algorithms produce improved optimization solutions in many kinds of optimization problems in several scientific sectors such as: engineering applications, computer science and electrical engineering problems, environmental forecasting, mechanical engineering, education, transportation, scheduling optimization, machine learning and deep learning model optimization.

Machine learning methods use training algorithms to create AI-based prediction models. According to numerous studies in the literature, swarm intelligence and evolutionary computation can be utilized with increased success in order to enhance the predictability and the generalization ability of the machine learning models.

One of the main advantages of metaheuristics in optimization problems is that they do not require gradient information and can be implemented on problems where the gradient cannot be acquired. Another advantage is that they can be applied on optimization problems where the objective or fitness function is unknown or when the objective function is not differentiable. Furthermore, metaheuristics can provide an optimal solution by using less computational power compared to other optimization algorithms.

There are several common characteristics among the metaheuristic models, such as the exploration and the exploitation processes. The exploration represents the capability of the optimization algorithm to investigate the search space globally by performing a wide search for feasible solutions, whereas exploitation represents the capability of improving the current solution by searching intensively in local regions for a better solution and it is generally performed at the last iterations of the algorithm.

The book is focused on presenting, discussing, categorizing and clarifying various aspects of swarm and evolutionary intelligence such as, boundary handling approaches, swarm topology variations, mutation, crossover and selection operators, termination conditions, hyper-parameter optimization, topology and weight optimization of machine learning and deep learning models.

The book comprises nine chapters. In addition, each chapter includes the most important modifications of the swarm and evolutionary algorithms, according to the literature, which were developed to improve their performance in several optimization problems. In Chapter 1, a theoretical introduction in computational optimization is performed regarding some basic concepts such as, minimization, maximization, constrained and unconstrained optimization, convex and non-convex optimization. Also, several important computational optimization techniques are discussed, regarding gradient and non-gradient-based methods and also some important machine learning training algorithms such as, Levenberg–Marquardt algorithm, scaled conjugate gradient, RMSProp, stochastic gradient descent. Furthermore, some significant new machine learning and deep learning optimizers are discussed such as: Adam optimizer, Adagrad and Adadelta. Chapter 2 is devoted to the theoretical foundation of evolutionary computation and genetic algorithm. The standard genetic algorithm and advanced aspects are discussed, such as:

mutation, crossover, selection operators and stopping criteria, bringing to the reader a complete aspect of the genetic algorithm and its latest operators. Furthermore, an important evolutionary algorithm is presented: the differential evolution and its variation, chaotic differential evolution. Also, an illustrative application example is presented to help the reader understand the differential evolution process. Chapter 3 presents the theoretical foundation of particle swarm optimization (PSO), the latest PSO algorithm advances, and illustrative examples. The standard bio-inspired particle swarm algorithm is explained and analyzed and some important algorithm variations such as: Quantum PSO, Chaotic PSO and mutation-based PSO algorithms. Advanced aspects and variants of particle swarm optimization are discussed, such as: swarm topologies, boundary approaches, swarm initialization techniques, stopping criteria and swarm mutation operators. Illustrative application examples are presented regarding unimodal, multimodal, convex and non-convex function approximation. Chapter 4 is devoted to the most important colony-based optimization algorithms: Ant colony optimization (ACO) and Artificial Bee Colony (ABC). Furthermore, several important ant-based algorithms are discussed such as, Ant System, Ant Colony System, Rank-based Ant System, Max-min Ant System and Population-based ACO. The standard Artificial Bee Algorithm is presented and also advanced aspects are discussed, such as: ABC selection methods, and boundary handling approaches of the artificial bee agent movement in the search space. Chapter 5 discusses the Cuckoo Search (CS) and the Bat Algorithm (BA). The standard Cuckoo Search based on the Lévy Flights is discussed and some important variations such as: Chaotic Cuckoo Search, Discrete Cuckoo Search and Discrete Binary Cuckoo Search. In addition, the Bat Algorithm and its variants are discussed including Chaotic Bat Algorithm, Binary Bat Algorithm, Discrete Bat Algorithm and also two recent variants: Bat Algorithm with double mutation (2020) and Self-adaptive bat algorithm (2019). Chapter 6 present and explain the swarm optimization algorithms: Firefly algorithm, Harmony Search and Cat Swarm Optimization algorithm. The standard algorithms are discussed and some important variants like firefly algorithm with Lévy flights, chaotic firefly algorithm, improved harmony search, chaotic harmony search, binary discreate cat swarm optimization, improved cat swarm optimization, chaotic bat algorithm, discreate bat algorithm and binary bat algorithm. Chapter 7 is devoted to the latest bio-inspired swarm algorithms such as: Grey Wolf Optimization

(GWO) Algorithm (2014), Whale Optimization Algorithm (WOA) (2016) and Grasshopper Optimization Algorithm (GOA) (2017). The original algorithms are discussed and also some important recent variants including Binary Grey Wolf Optimization (2016), Grey Wolf with Lévy flight (2017), Whale Optimization with Lévy Flight (2017), Binary Whale Optimization Algorithm (2020), Improved Grasshopper Optimization Algorithm (2018) and Chaotic Grasshopper Optimization (2019). In Chapter 8 machine learning optimization applications are illustrated focusing on artificial neural network optimization. A neural network can be optimized by using swarm and evolutionary optimization methods in several ways, such as: weight optimization, structure optimization, or feature selection. Illustrative real-world comparative examples are presented by using genetic algorithm, particle swarm optimization and ant colony optimization. Chapter 9 is devoted to deep learning topology and weight optimization by utilizing swarm intelligence and evolutionary computation, including deep unidirectional and bi-directional long short-term memory (LSTM and Bi-LSTM) network optimization and also deep Convolutional Neural Networks (CNN) optimization. The experimental part includes deep learning optimization applications such as covid19 diagnosis from chest x-ray images.

The book presents some state-of-the-art research involving swarm intelligence and evolutionary algorithms together with their variants with pseudo-codes aiming at delivering a more robust and comprehensive knowledge of the recent advances and its applications in machine learning and deep learning.

Dr. Georgios N. Kouziokas
University of Thessaly
Greece

Contents

Contents

Cı ıAPTER 1

Computational Optimization

◇◇

In this chapter a theoretical introduction of computational optimization is discussed regarding minimization, maximization, constrained and unconstrained optimization, convex and non-convex optimization. Also, several important computational optimization techniques are discussed, such as: Gauss-Newton and quasi-Newton methods, gradient-based methods such as steepest descent, conjugate gradient, and also non-gradient methods such as genetic algorithm and swarm intelligence algorithms. Also, several important optimizers used in machine learning are presented such as: Levenberg–Marquardt algorithm, Scaled Conjugate Gradient, RMSProp, Stochastic Gradient Descent. Furthermore, several important and relatively new optimizers are discussed such as: Adam optimizer (2014), which is an important optimizer, frequently used in machine learning and deep learning, Adagrad (2012) and Adadelta (2012).

Computational optimization

Introduction

A computational optimization problem can be formulated as a minimization or maximization problem. Let us consider a computational minimization problem:

$$minimize \ f_i(x), \ (i \ = \ 1,2,...,M) \tag{1.1}$$

subject to the following constraints:

$$h_j(x) = c_j, \ \text{for} \ j \ = \ 1,2,...,J \tag{1.2}$$

$$g_k(x) \leq d_k, \text{ for } k = 1,2,...,K \tag{1.3}$$

$$x = (x_1, x_2, ..., x_n) \tag{1.4}$$

where, $h_j(x) = c_j$, for j = 1, 2, ..., J and $g_k(x) \leq d_k$, for k = 1, 2, ..., K are constraints that must be satisfied. The functions $f_i(x)$, $h_j(x)$ and $g_k(x)$ are nonlinear in general. $h_j(x)$ represents the equality constraints of an optimization problem, $g_k(x)$ represents the inequality constraints of an optimization problem.

The vector x can be either continuous or discrete or mixed in a multi-dimensional search space. The $f_i(x)$ functions are nominated as objective or fitness or cost functions. When the parameter $M > 1$, then we deal with a multi-objective optimization problem. Every minimization problem can be expressed as a maximization problem by substituting $f_i(x)$ with $-f_i(x)$.

When the functions $f_i(x)$, $h_j(x)$ and $g_k(x)$ are linear then linear programming techniques can be used. When the variables of an optimization problem take only discrete values then we have a discrete optimization problem. When the variables of an optimization problem are continuous then we have to solve a continuous optimization problem. When some variables are continuous, and some other discrete then the optimization problem is mixed. When an objective function to be minimized is convex then we have a convex optimization problem which has a globally optimal solution (Koziel and Yang, 2011).

Optimization methods

Gauss-Newton method

Gauss-Newton method can be derived from a second order Taylor expansion of the fitness function to define the search direction, expressed by the equation:

$$p = -H_e^{-1}g \tag{1.5}$$

where H_e represents the exact Hessian matrix from the second derivatives and g represents the gradient vector. The gradient vector is calculated by the equation:

$$g = J^H (u_c(\theta) - u_m) \tag{1.6}$$

The Hessian matrix is calculated by the equation:

$$H = J^H J \qquad (1.7)$$

where,

- $J - \partial u_c / \partial \theta$ represents the Jacobian matrix with dimensions $3n \times M$
- $3n$ represents the number of entries in u_c
- M represents the number of unknowns in θ

The i-th column in the Jacobian matrix is calculated by the equation:

$$J_i = K^{-1} \left[\frac{\partial K}{\partial \theta_i} u_c \right] \qquad (1.8)$$

where K represents the stiffness matrix. The Gauss-Newton method converges faster than the gradient-based optimization algorithms because it utilizes second derivative information, but it has an increased computational cost because of the Jacobian matrix formation and the Hessian matrix factorization.

$$H_r = H + \gamma_r I \qquad (1.9)$$

The error vector is expressed by the equation:

$$e = [e_{1,1} e_{1,2} \cdots e_{1,m} \cdots e_{p,1} \cdots e_{p,m}]^T \qquad (1.10)$$

The update rule for the Gauss-Newton algorithm can be expressed by the following equation:

$$W_{k+1} = W_k - (J_k^T J_k)^{-1} J_k e_k \qquad (1.11)$$

Quasi-Newton method

The Quasi-Newton method utilizes an update approximation of Hessian matrix in each iteration. The equation that calculates the inversion of the Hessian matrix is expressed by the following equation:

$$H_{k+1}^{-1} = H_k^{-1} + \frac{(s_k^T y_k + y_k^T H_k^{-1} y_k)(s_k s_k^T)}{(s_k^T y_k)^2} - \frac{H_k^{-1} y_k s_k^T + s_k y_k^T H_k^{-1}}{s_k^T y_k} \qquad (1.12)$$

where,

- $s_k = \theta_k - \theta_{k-1}$
- $y_k = g_k - g_{k-1}$

3

The inverted Hessian can be utilized to calculate the search direction:

$$\boldsymbol{p}_k = -\boldsymbol{H}_{k+1}^{-1}\boldsymbol{g}_k \tag{1.13}$$

Gradient-based optimization

Let us consider a function $f(x)$. The gradient of this function is defined by the partial derivatives vector for each of the independent variables:

$$\nabla f(x) \equiv g(x) \equiv \begin{bmatrix} \dfrac{\partial f}{\partial x_1} \\[2mm] \dfrac{\partial f}{\partial x_2} \\[2mm] ... \\[2mm] \dfrac{\partial f}{\partial x_n} \end{bmatrix} \tag{1.14}$$

The gradient of a function with n variables is a n-dimensional vector. The second derivative of a n dimensional function is determined by n^2 partial derivatives:

$$\frac{\partial^2 f}{\partial x_i \partial x_j}, \text{ with } i \neq j \text{ and } \frac{\partial^2 f}{\partial x_i^2} \text{ with } i = j \tag{1.15}$$

The partial derivatives are expressed as:

$$\frac{\partial f}{\partial x_i}, \frac{\partial f}{\partial x_j}, \frac{\partial^2 f}{\partial x_i \partial x_j} \tag{1.16}$$

If these partial derivatives are continuous and the *f function* is single valued, then:

$$\frac{\partial^2 f}{\partial x_i \partial x_j} = \frac{\partial^2 f}{\partial x_j \partial x_i} \tag{1.17}$$

4

Consequently, the second-order partial derivatives can be expressed by a square symmetric matrix called the *Hessian matrix*:

$$H(x) = \begin{bmatrix} \dfrac{\partial^2 f}{\partial x_1^2} & \dfrac{\partial^2 f}{\partial x_1 \partial x_2} & \cdots & \dfrac{\partial^2 f}{\partial x_1 \partial x_N} \\[2mm] \dfrac{\partial^2 f}{\partial x_2 \partial x_1} & \dfrac{\partial^2 f}{\partial x_2^2} & \cdots & \dfrac{\partial^2 f}{\partial x_2 \partial x_N} \\[2mm] \cdots & \cdots & \cdots & \cdots \\[2mm] \dfrac{\partial^2 f}{\partial x_N \partial x_1} & \dfrac{\partial^2 f}{\partial x_N \partial x_2} & \cdots & \dfrac{\partial^2 f}{\partial x_N^2} \end{bmatrix} \quad (1.18)$$

which contains n (n + 1)/2 elements.

Steepest or gradient descent algorithm

In the Steepest Descent Algorithm, also known as gradient Descent Algorithm proposed by Cauchy (Cauchy, 1847) let us consider a step size learning constant α (learning rate), an N-dimensional vector:

$$w = \left[w_1, \ldots, w_N \right]^T \quad (1.19)$$

The N-dimensional vector represents the parameters, and $E(x, w)$ the objective function. The gradient can be calculated by the equation:

$$g = \frac{\partial E(x, w)}{\partial w} = \left[\frac{\partial E}{\partial w_1} \frac{\partial E}{\partial w_2} \cdots \frac{\partial E}{\partial w_N} \right]^T \quad (1.20)$$

The gradient-descent update rule of the steepest descent algorithm can be calculated from the gradient (g) by using the following equation:

$$w_{k+1} = w_k - \alpha g_k \quad (1.21)$$

where, α represents the step size learning constant.

Conjugate gradient algorithms

The conjugate gradient method (Fletcher, 1976) is a modification of the steepest descent method. Steepest Descent suffers from very slow convergence. During the optimization the conjugate vectors are generating but a new vector p_k is computed only by using the previous p_{k-1} vectors, and not all the previous vectors of the conjugate set. Due to this mechanism the method requires less storage and computation.

Conjugate gradient method improves the steepest descent by developing a set of search directions as the following equation shows:

$$\boldsymbol{p}_k = -\boldsymbol{g}_k + \beta_k \boldsymbol{p}_{k-1} \tag{1.22}$$

where,

- p_k represents the current direction and p_{k-1} the previous direction
- g_k represents the current residual parameter
- β_k is given by the Polak-Ribére formula (Polak and Ribiere, 1969) or the Hestenes–Stiefel formula, or the Fletcher–Reeves formula:

The Hestenes–Stiefel formula (Hestenes and Stiefel, 1952) is given by the equation:

$$\beta_k^{HS} = \frac{g_k^T y_k}{d_{k-1}^T y_k} \tag{1.23}$$

The Fletcher–Reeves formula (Fletcher and Reeves, 1964) is given by the equation:

$$\beta_k^{FR} = \frac{g_k^T g_k}{g_{k-1}^T g_{k-1}} \tag{1.24}$$

The Polak–Ribière formula (Polak and Ribiere, 1969) is given by the equation:

$$\beta_k^{PR} = \frac{g_k^T y_k}{g_{k-1}^T g_{k-1}} \tag{1.25}$$

where:

$$y_k = g_k - g_{k-1} \tag{1.26}$$

The Conjugate gradient method can be more efficient compared to the Gauss-Newton method. The Conjugate gradient method requires less memory storage resources because of the simple vector computations.

Conjugate gradient algorithms are considered as optimization methods between the steepest descent and Newton methods. Another advantage is that there is no need to store any matrices as in the Newton method, and in the quasi-Newton methods, and also, they converge faster compared to the steepest descent method.

Optimizers for machine learning

Stochastic gradient descent

Stochastic gradient descent is an efficient optimization algorithm especially when the dataset is very large because of the low computational cost in each iteration. Several applications of stochastic gradient descent are studied in the literature in several scientific areas such as machine learning and signal processing (Murphy, 2012). Stochastic gradient descent estimates a gradient in each iteration on a randomly selected sample (mini-batch) and updates the model variable.

In machine learning the conventional gradient descent algorithm updates the weights and biases in order to minimize the selected loss function and is calculated by using the negative gradient of the loss function:

$$\theta_{\ell+1} = \theta_\ell - \alpha \nabla E(\theta_\ell) \tag{1.27}$$

where,

- ℓ represents the iteration number
- $\alpha > 0$ represents the learning rate
- θ represents the vector parameter,
- $E(\theta)$ represents the loss function.

In the conventional gradient descent algorithm, the loss function gradient $\nabla E(\theta)$, is calculated by using the whole training set of the dataset at once, but the *stochastic* gradient descent algorithm calculates the loss function gradient and updates the parameters by using only a subset (mini-batch) of the training data at each iteration.

Stochastic gradient descent with momentum

One disadvantage of the stochastic gradient descent algorithm is that it can oscillate in the path of the gradient descent in the direction of the optimum. The Stochastic Gradient Descent with Momentum uses a momentum parameter in the parameter update equation in order to reduce the oscillation (Murphy, 2012). The stochastic gradient descent with momentum update equation is expressed as:

$$\theta_{\ell+1} = \theta_\ell - \alpha \nabla E(\theta_\ell) + \gamma(\theta_\ell - \theta_{\ell-1}) \tag{1.28}$$

where,

- $\gamma \in [0,1)$ represents the momentum of the previous iteration of the gradient to the current iteration
- ℓ represents the iteration number
- $\alpha > 0$ represents the learning rate
- θ represents the vector parameter,
- $E(\theta)$ represents the loss function.

Levenberg–Marquardt algorithm

The Levenberg-Marquardt algorithm is considered as one of the fastest supervised back propagation algorithms (Marquardt, 1963). Levenberg–Marquardt algorithm is used as a technique to solve non-linear problems. Levenberg–Marquardt algorithm is a combination of the steepest descent method and the Newton algorithm. The Levenberg–Marquardt algorithm update rule is expressed by the following equation:

$$\Delta w = (J^T J + \mu I)^{-1} J^T e \tag{1.29}$$

where,

- w represents the weight vector
- I represents the identity matrix
- μ represents the combination coefficient
- J represents the Jacobian matrix
- e represents the error vector

The dimensions of the Jacobian matrix are (P × M) × N and the dimensions of the error vector are (P × M) × 1 and they are calculated by the following equations:

$$
J = \begin{bmatrix}
\dfrac{\partial e_{1,1}}{\partial w_1} & \dfrac{\partial e_{1,1}}{\partial w_2} & & \dfrac{\partial e_{1,1}}{\partial w_N} \\[2ex]
\dfrac{\partial e_{1,2}}{\partial w_1} & \dfrac{\partial e_{1,2}}{\partial w_2} & \cdots & \dfrac{\partial e_{1,2}}{\partial w_N} \\[2ex]
\cdots & \cdots & \cdots & \cdots \\[2ex]
\dfrac{\partial e_{1,M}}{\partial w_1} & \dfrac{\partial e_{1,M}}{\partial w_2} & \cdots & \dfrac{\partial e_{1,M}}{\partial w_N} \\[2ex]
\cdots & \cdots & \cdots & \cdots \\[2ex]
\dfrac{\partial e_{P,1}}{\partial w_1} & \dfrac{\partial e_{P,1}}{\partial w_2} & \cdots & \dfrac{\partial e_{P,1}}{\partial w_N} \\[2ex]
\dfrac{\partial e_{P,2}}{\partial w_1} & \dfrac{\partial e_{P,2}}{\partial w_2} & \cdots & \dfrac{\partial e_{P,2}}{\partial w_N} \\[2ex]
\cdots & \cdots & \cdots & \cdots \\[2ex]
\dfrac{\partial e_{P,M}}{\partial w_1} & \dfrac{\partial e_{P,M}}{\partial w_2} & \cdots & \dfrac{\partial e_{P,M}}{\partial w_N}
\end{bmatrix}
\tag{1.30}
$$

$$
e = \begin{bmatrix}
e_{1,1} \\
e_{1,2} \\
\cdots \\
e_{1,M} \\
\cdots \\
e_{P,1} \\
e_{P,2} \\
\cdots \\
e_{P,M}
\end{bmatrix}
\tag{1.31}
$$

where,
- *P* represents the number of training patterns
- *M* represents the number of outputs
- *N* represents the number of weights.

9

The error vector e is estimated by the following equation:

$$e_{p,m} = d_{p,m} - o_{p,m}$$ (1.32)

where,

- $d_{p,m}$ represents the desired (real) output
- $o_{p,m}$ represents the output vector
- m represents the neural network outputs
- p represents the training patterns.

During the algorithm process, the Jacobian matrix J is estimated and stored and then the multiplications of the Jacobian matrix are calculated to update the weights by using the equation (1.29). For large sized problems, there is a memory storage limitation problem for the Jacobian matrix J.

Wilamowski and Yu (2010), proposed an improved computation method for Levenberg–Marquardt used to optimize the neural network training process, where the gradient vector and Quasi-Hessian matrix are calculated without using the Jacobian matrix multiplication, as shown below.

The sum of squared errors (SSE) is utilized to evaluate the training process as expressed with the equation:

$$E(w) = \frac{1}{2} \sum_{p=1}^{P} \sum_{m=1}^{M} e_{p,m}^2$$ (1.33)

where $e_{p,m}$ represents the error of output m under a training pattern p, defined by the equation (1.32).

The Hessian matrix H with dimensions $N \times N$ is expressed by the following:

$$H = \begin{bmatrix} \dfrac{\partial^2 E}{\partial w_1^2} & \dfrac{\partial^2 E}{\partial w_1 \partial w_2} & \cdots & \dfrac{\partial^2 E}{\partial w_1 \partial w_N} \\ \dfrac{\partial^2 E}{\partial w_2 \partial w_1} & \dfrac{\partial^2 E}{\partial w_2^2} & \cdots & \dfrac{\partial^2 E}{\partial w_2 \partial w_N} \\ \cdots & \cdots & \cdots & \cdots \\ \dfrac{\partial^2 E}{\partial w_N \partial w_1} & \dfrac{\partial^2 E}{\partial w_N \partial w_2} & \cdots & \dfrac{\partial^2 E}{\partial w_N^2} \end{bmatrix}$$ (1.34)

where N represents the number of weights.

By combining the two previously mentioned equations the Hessian matrix **H** can be calculated by the following equation:

$$\frac{\partial^2 E}{\partial w_i \partial w_j} = \sum_{p=1}^{P} \sum_{m=1}^{M} \left(\frac{\partial e_{p,m}}{\partial w_i} \frac{\partial e_{p,m}}{\partial w_j} + \frac{\partial^2 e_{p,m}}{\partial w_i \partial w_j} e_{p,m} \right) \qquad (1.35)$$

where i and j represent the weight indexes.

The previous equation of Levenberg–Marquardt algorithm can be approximated (Hagan and Menhaj, 1994; Peng et al., 2008) as:

$$\frac{\partial^2 E}{\partial w_i \partial w_j} \approx \sum_{p=1}^{P} \sum_{m=1}^{M} \left(\frac{\partial e_{p,m}}{\partial w_i} \frac{\partial e_{p,m}}{\partial w_j} \right) = q_{ij} \qquad (1.36)$$

where q_{ij} represents the quasi-Hessian matrix element with a row i and a column j.

By combining the previous equation with equations (1.30) and (1.31), the quasi-Hessian matrix **Q** can be estimated by using the Hessian matrix with the following equation:

$$H \approx Q = J^T J \qquad (1.37)$$

The gradient vector **g** can be calculated by the equation:

$$g = \left[\frac{\partial E}{\partial w_1} \quad \frac{\partial E}{\partial w_2} \quad \cdots \quad \frac{\partial E}{\partial w_N} \right]^T \qquad (1.38)$$

Each element of the gradient vector g can be estimated by the equation:

$$g_i = \frac{\partial E}{\partial w_i} = \sum_{p=1}^{P} \sum_{m=1}^{M} \left(\frac{\partial e_{p,m}}{\partial w_i} e_{p,m} \right) \qquad (1.39)$$

Consequently, the gradient vector **g** can be expressed by using the Jacobian matrix **J** according to the equation:

$$g = J^T e \qquad (1.40)$$

The Levenberg–Marquardt algorithm update rule can be expressed by the following equation:

$$\Delta w = (Q + \mu I)^{-1} g \qquad (1.41)$$

Wilamowski and Yu (2010), showed that the last equation produces the same results with the initial equation (1.29) of the Levenberg–Marquardt algorithm. The difference in the proposed method is that the quasi-Hessian matrix and the gradient vector are estimated directly without calculating and storing the Jacobian matrix as in equations (1.30) and (1.31).

The authors also proposed an improved computation of the quasi-Hessian matrix by introducing a quasi-Hessian submatrix q_{pm}. The quasi-Hessian matrix Q can be estimated as the sum of all the submatrices q_{pm}, with the following equation:

$$Q = \sum_{p=1}^{P}\sum_{m=1}^{M} q_{p,m} \tag{1.42}$$

Furthermore, Wilamowski and Yu (2010), proposed an improved calculation of the gradient vector by introducing a Gradient sub vector η_{pm} with dimensions: $N \times 1$. The gradient vector g can be estimated as the sum of all the gradient sub vectors η_{pm} with the following equation:

$$g = \sum_{p=1}^{P}\sum_{m=1}^{M} \eta_{p,m} \tag{1.43}$$

where η_{pm} is calculated by the following equation:

$$\eta_{pm} = \begin{bmatrix} \dfrac{\partial e_{p,m}}{\partial w_1} e_{pm} \\ \dfrac{\partial e_{p,m}}{\partial w_2} e_{pm} \\ \cdots \\ \dfrac{\partial e_{p,m}}{\partial w_N} e_{p,m} \end{bmatrix} = \begin{bmatrix} \dfrac{\partial e_{p,m}}{\partial w_1} \\ \dfrac{\partial e_{p,m}}{\partial w_2} \\ \cdots \\ \dfrac{\partial e_{p,m}}{\partial w_N} \end{bmatrix} \times e_{p,m} \tag{1.44}$$

Scaled conjugate gradient algorithm

A Scaled Conjugate Gradient Algorithm was proposed in 1993 to deal with large-scale optimization problems in a more effective way (Moller, 1993). The Scaled Conjugate Gradient (SCG) algorithm is used as an optimizer in the neural network training process which requires less memory compared to other optimizers such as Levenberg Marquardt and the Bayesian Regularization (BR) algorithms (Møller, 1993). Scaled Conjugate Gradient Algorithm uses second order information in neural network training, and has low memory requirements because of the reduced complexity of the gradient calculations (Møller, 1993).

The steps of the SCG algorithm are summarized below (Møller, 1993):

1. Choose the initial weight vector, and set the values of scalars σ, λ_1, and $\overline{\lambda}_k$.

   Set the initial solution p_1, and the direction of the steepest descent r_1, equal to the estimated error surface gradient: $p_1 = r_1 = -\nabla f(x_1)$ and $k = 1$.

   Set *success* = *true*.

2. If *success* = *true*, then calculate the second order information, $\overline{s}_k$:

$$\sigma_k = \frac{\sigma}{|p_k|} \tag{1.45}$$

$$\overline{s}_k = \frac{\nabla f(x_k + \sigma_k p_k) - \nabla f(x_k)}{\sigma_k} \tag{1.46}$$

$$\delta_k = p_k^T \overline{s}_k \tag{1.47}$$

3. Scale δ_k:

$$\delta_k = \delta_k + (\lambda_k - \overline{\lambda}_k)|p_k|^2 \tag{1.48}$$

4. If $\delta_k \leq 0$, then make the Hessian matrix positive definite:

$$\overline{\lambda}_k = 2\left(\lambda_k - \frac{\delta_k}{|p_k|^2}\right) \tag{1.49}$$

$$\delta_k = -\delta_k + \lambda_k|p_k|^2 \tag{1.50}$$

$$\lambda_k = \overline{\lambda}_k \tag{1.51}$$

5. Estimate the μ_k step size:

$$\mu_k = p_k^T r_k \tag{1.52}$$

$$\alpha_k = \frac{\mu_k}{\delta_k} \tag{1.53}$$

6. Estimate the Δ_k comparison parameter:

$$\Delta_k = \frac{2\delta_k(f(x_k) - f(x_k + \alpha_k p_k))}{\mu_x^2} \tag{1.54}$$

7. If $\Delta_k \geq 0$, then a successful reduction in error can be made:

Update the weight vectors, x_{k+1}, and the steepest descent direction, r_{k+1}.

$\overline{\lambda}_k = 0$, *success* = *True*

if $k \bmod N = 0$ then restart the algorithm

$$p_{k+1} = r_{k+1} \tag{1.55}$$

else, create a new conjugate direction:

$$\beta_k = \frac{|r_{k+1}|^2 - r_{k+1}^T r_k}{\mu_k} \tag{1.56}$$

$$p_{k+1} = r_{k+1} + \beta_k p_k \tag{1.57}$$

If $\Delta_k \geq 0.75$ then reduce the scale parameter $\lambda_k = (1/4)\lambda_k$,

else

$\overline{\lambda}_k = \lambda_k$, success = false

8. If $\Delta_k < 0.25$, then increase the scale parameter

$$\lambda_k = \lambda_k + (\delta_k(1 - \Delta_k)/|p_k|^2) \tag{1.58}$$

9. If the steepest descent direction is $r_k \neq 0$, then set $k = k+1$ and go to step 2. Else, terminate optimization and return x_{k+1} as the desired minimum.

Adagrad

The Adagrad is a gradient-based optimization algorithm which adjusts the learning rate for each parameter (Dean et al., 2012). Adagrad implements a different learning rate for each parameter in a time step. Let g_i^t represent the gradient at a time step (iteration) t. Then for the steps (iterations) $0,1,...,$ $t-1$ the corresponding gradients are $g_i^0, g_i^1,...,g_i^s$. The sum of squares G of the gradients is calculated by the equation:

$$G = \left(g_i^0\right)^2 + \left(g_i^1\right)^2 + \cdots + \left(g_i^{t-1}\right)^2 \tag{1.59}$$

The update rule in Adagrad adjusts the global learning rate at each time step for every parameter from the past gradients based on the equation:

$$x_i = x_i - \frac{\alpha}{G^{\frac{1}{2}}} g_i^t \tag{1.60}$$

where,

- α represents the global learning rate, which is adjusted for each parameter.
- x_i denotes the parameter

Adagrad algorithm improves the robustness of SGD and can be applied with success on large-scale datasets for training neural networks (Dean et al., 2012; Duchi et al., 2011).

RMSProp

The stochastic gradient descent with momentum implements a single learning rate factor for all parameters. Several optimization algorithms use different learning rates for every parameter automatically adapting the loss function in order to improve training. In a similar way, RMSProp (Initials from root mean square propagation) implements a moving average of the squares of the parameter gradients (Riedmiller and Braun, 1993):

$$v_\ell = \beta_2 v_{\ell-1} + (1 - \beta_2)[\nabla E(\theta_\ell)]^2 \qquad (1.61)$$

where β_2 represents the decay rate of the moving average.

The RMSProp algorithm implements the moving average in the updates rule for each parameter according to the equation,

$$\theta_{\ell+1} = \theta_\ell - \frac{\alpha \nabla E(\theta_\ell)}{\sqrt{v_\ell} + \epsilon} \qquad (1.62)$$

where,

• ε is a small constant in order to avoid division by zero.

The RMSProp algorithm decreases the learning rates when the parameters have large gradients and increases the learning rates when the parameters have small gradients.

Adadelta

Zeiler (2012), proposed the Adadelta algorithm, in order to improve the two drawbacks of the Adagrad optimizer: the continual decay of the individual learning rates during training, and the need for choosing a learning rate globally. In the Adagrad optimizer the squared gradients for every iteration are accumulated and consequently, the learning rate is shrinking on every dimension and finally becomes very small. In order to deal with these problems, the proposed main update rule for the Adadelta algorithm is expressed by the following equation (Zeiler, 2012):

$$x_i = x_i - \frac{RMS\left[\Delta x_i\right]^{t-1}}{RMS\left[g_i\right]^t} g_i^t \tag{1.63}$$

where,

- $RMS\left[\Delta x\right]^{t-1}$ represents the root-mean-square updates of x.
- t represents the time step.

Adam optimizer

Adam optimizer is an algorithm that uses an adaptive learning rate (Kingma and Ba, 2014). Adam optimizer has been applied in several deep learning techniques. During the algorithm process it finds single adaptive learning rates for the individual parameters. The algorithm name is derived by the adaptive moment calculation process of the algorithm. The first and second moments are the mean and the variance, accordingly. Adam optimizer implements exponentially moving averages for moment calculation in each mini batch for every iteration. The update rules for Adam optimizer gradient moving averages and the squared gradient accordingly, are expressed by the following equations:

$$m_t = \beta_1 m_{t-1} + (1 - \beta_1) g_t \tag{1.64}$$

$$v_t = \beta_2 v_{t-1} + (1 - \beta_2) g_t^2 \tag{1.65}$$

where,

- m and v are the moving averages of the gradient and the squared gradient accordingly.
- g represents the gradient in the current batch
- t represents the number of iterations
- β_1 and β_2 represent the hyper-parameters of the algorithm regarding the gradient and the squared gradient accordingly.

The Bias-corrected calculators for the first and second moments are given by the following equations:

$$\hat{m}_t = \frac{m_t}{1 - \beta_1^t} \tag{1.66}$$

16

$$\hat{v}_t = \frac{v_t}{1 - \beta_2^t}$$

(1.67)

At the final stage of the algorithm the moving averages are scaled to the individual learning rates for every parameter and the weight update is computed by the following equation:

$$w_t = w_{t-1} - \alpha \frac{\hat{m}_t}{\sqrt{\hat{v}_t} + \epsilon}$$

(1.68)

where,

- w represents weight of every model.
- α represents learning rate.
- t represents the number of the iteration.
- ε represents a parameter used to prevent division by zero.

Non-gradient methods

The gradient-based algorithms are efficient in several optimization problems: convex problems, non-linearly constrained problems, but the most gradient-based algorithms have problems in optimizing discontinuous, multi-modal, mixed discrete-continuous functions. Also, gradient-based algorithms have problems in multiple local minima, or when an objective function is non-differentiable.

Several gradient-free or non-gradient methods are based on the natural behavior of animals or insects or use heuristic methods. These methods include: genetic algorithms, evolution strategies, heuristic methods and several kinds of swarm intelligence methods that will be discussed in this book such as: Particle swarm optimization, Firefly algorithm, Harmony search, Cat swarm optimization, Grey Wolf Optimization Algorithm, Whale Optimization, Grasshopper Optimization, Ant colony optimization and Artificial Bee Colony, Cuckoo search and Bat algorithm.

References

Cauchy, A. (1847). Méthode générale pour la résolution des systemes d'équations simultanées. *Comp. Rend. Sci. Paris*, 25(1847): 536–538.

Dean, J., Corrado, G., Monga, R., Chen, K., Devin, M. et al. (2012). Large scale distributed deep networks. pp. 1232–1240. *In*: *Conf. on Neural Information Processing Systems (NeurIPS)*.

Duchi, J., Hazan, E. and Singer, Y. (2011). Adaptive subgradient methods for online learning and stochastic optimization. *Journal of Machine Learning Research*, 12(7).

Fletcher, R. and Reeves, C. M. (1964). Function minimization by conjugate gradients. *The Computer Journal*, 7(2): 149–154.

Fletcher, R. (1976). Conjugate gradient methods for indefinite systems. pp. 73–89. *In*: *Numerical Analysis*. Springer, Berlin, Heidelberg.

Hagan, M. T. and Menhaj, M. B. (1994). Training feedforward networks with the Marquardt algorithm. *IEEE Transactions on Neural Networks*, 5(6): 989–993.

Hestenes, M. R. and Stiefel, E. (1952). *Methods of Conjugate Gradients for Solving Linear Systems* (Vol. 49, No. 1). Washington, DC: NBS.

Kingma, D. P. and Ba, J. (2014). Adam: A method for stochastic optimization. *arXiv preprint arXiv:1412.6980*.

Koziel, S. and Yang, X. S. (eds.). (2011). *Computational Optimization, Methods and Algorithms* (Vol. 356). Springer.

Marquardt, D. W. (1963). An algorithm for least-squares estimation of nonlinear parameters. *Journal of the Society for Industrial and Applied Mathematics*, 11: 431–441.

Møller, M. F. (1993). A scaled conjugate gradient algorithm for fast supervised learning. *Neural Networks*, 6(4): 525–533.

Murphy, K. P. (2012). Machine Learning: A Probabilistic Perspective. MIT press.

Peng, J. X., Li, K. and Irwin, G. W. (2008). A new Jacobian matrix for optimal learning of single-layer neural networks. *IEEE Transactions on Neural Networks*, 19(1): 119–129.

Polak, E. and Ribiere, G. (1969). Note sur la convergence de méthodes de directions conjuguées. *ESAIM: Mathematical Modelling and Numerical Analysis-Modélisation Mathématique et Analyse Numérique*, 3(R1): 35–43.

Riedmiller, M. and Braun, H. (1993, March). A direct adaptive method for faster backpropagation learning: The RPROP algorithm. pp. 586–591. *In*: *IEEE International Conference on Neural Networks*. IEEE.

Wilamowski, B. M. and Yu, H. (2010). Improved computation for Levenberg–Marquardt training. *IEEE Transactions on Neural Networks*, 21(6): 930–937. doi:10.1109/TNN.2010.2045657.

Zeiler, M. D. (2012). Adadelta: an adaptive learning rate method. *arXiv preprint arXiv:1212.5701*.

CHAPTER 2

Evolutionary Computation and Genetic Algorithm

This chapter is devoted to the theoretical foundation of evolutionary computation and genetic algorithm. The standard genetic algorithm is presented and discussed. Advanced aspects of genetic algorithm are discussed, such as: mutation, crossover and selection operators and stopping criteria, bringing to the reader a complete aspect of the genetic algorithm and its latest operators. Furthermore, an important evolutionary algorithm is presented: the differential evolution and its variation: chaotic differential evolution. Also, an illustrative application example is presented to help the reader understand the differential evolution process.

Evolutionary strategy

Evolutionary strategy algorithm proposed to be used in technical optimization problems where there is no specific objective function and no other conventional optimization method could be applied (Alavi and Henderson, 1981; Rechenberg, 1965; Schwefel, 1981). Compared to genetic algorithm, in the evolutionary strategy there is only a mutation operator.

The main steps in the evolutionary strategy algorithm are described below (Negnevitsky, 2005):

Step 1: Select the number of variables N to idealize the problem, and then define the range for every variable.

Step 2: Choose randomly a set of these variables for the initialization of the parent population.

Step 3: Estimate the solutions corresponding to these parent variables.

Step 4: Produce a new (offspring) variable by adding a normally distributed random variable with zero mean and pre-selected deviation to every parent variable, which reflect the natural evolution process with smaller changes.

Step 5: Estimate the solution related to the offspring variables.

Step 6: Compare the solution related to the offspring variables with the one related to the parent variables. If the offspring solution is better than the parent solution, then substitute the parent population with the population of the offsprings. Otherwise, maintain the parent variables.

Step 7: Go to the 4th step, and repeat the procedure until the termination criterion is reached, or until the maximum number of generations or a desirable value of the solution is attained.

Evolution strategies can be used to solve several kinds of optimization problems and produce improved results compared to many conventional optimization techniques (Back, 1996; Schwefel, 1995). The basic difference of evolution strategy compared to the genetic algorithm is that it utilizes only mutation, while the genetic algorithm utilizes both crossover and mutation.

Genetic algorithm

Genetic algorithm is a population-based algorithm similar to other evolutionary algorithms. The main operators of genetic algorithm are: selection, crossover, and mutation. Genetic algorithm proposed by John Holland (1975) in the book "Adaptation in Natural and Artificial System" which discusses the principles in natural evolution and how they can be applied on optimization problems. Genetic algorithm was used in several optimization problems such as vehicle routing problem (Baker and Ayechew, 2003; Berger and Barkaoui, 2004; May et al.) autonomous mobile robot path planning (Lamini et al., 2018), economic load dispatch (Sahay, 2018), and optimization in wireless sensor networks (Ferentinos and Tsiligiridis, 2007; Rodriguez et al., 2015).

The main steps in the genetic algorithm are described below (Negnevitsky, 2005):

Step 1: Idealize the problem as a chromosome problem, select the size of a chromosome population, the probability of the crossover and the mutation.

Step 2: Determine a fitness function.

Step 3: Generate randomly the initial chromosome population N.

Step 4: Estimate the fitness function value of each chromosome.

Step 5: Choose a pair of chromosomes for mating from the population. The parent chromosomes are chosen with a probability according to their fitness. The increased fit chromosomes have an increased probability for being selected for mating than the other chromosomes.

Step 6: Create a pair of children (offspring) chromosomes by implementing crossover and mutation operators.

Step 7: Put the new children (offspring) chromosomes in the new chromosome population.

Step 8: Repeat Step 5 until the new population size of chromosomes is equal to the initial population size.

Step 9: Replace the initial population of chromosomes with the new children (offspring) population.

Step 10: Go to Step 4, and repeat until the termination criterion is met.

The algorithm includes three basic parameters:

Selection: A selection process of the best individuals for reproduction based on their fitness value.

Crossover: Mechanism of individual genetic data merging.

Mutation: The mutation mechanism is based on the alteration of a percentage (mutation rate) of the genes in a chromosome.

Initialization

The genetic algorithm begins with a random population. Each chromosome has a set of parameters simulating the genes. The main scope of initialization is to sparse the positions (solutions) in the search space uniformly to increase the population diversity.

Selection methods

The main genetic algorithm utilizes a roulette wheel method to set the probabilities to the individuals and select them for the next generation according to their fitness values. Several selection methods are presented in the next sections.

Tournament selection

Tournament selection is the most utilized selection method for the Genetic Algorithm. In this selection, a set of 'n' individuals are selected from the population in a random way. The individual with the best fitness value is selected for the next stage of the genetic algorithm (Goldberg and Deb, 1991). The main advantages of this selection method are: less algorithm process time and easy implementation.

Linear ranking selection

In the linear ranking selection technique (Baker, 1985), individuals are ranked according to their fitness value and then the rank value is assigned to each one. The best individual is the one with rank 'N' and the worst one with rank '1'. The linear selection probability is expressed by the equation:

$$P_i = \frac{1}{N}(n^- + (n^+ - n^-)\frac{i-1}{N-1}); i \in \{1, \ldots, N\} \qquad (2.1)$$

where,

- P_i represents the selection probability of the i_{th} individual.

- $\dfrac{n^-}{N}$ represents the selection probability of the worst individual solution

- $\dfrac{n^+}{N}$ represents the selection probability of the best individual solution.

Proportionate roulette wheel selection

When the selection probability is proportional in the roulette wheel regarding each sector of a partitioned wheel then it is called proportionate roulette wheel selection.

The proportion of individual fitness values against the total fitness values (solutions) of the population defines the probability of selection in the next generation. This consequently decides the area occupied by that individual on the wheel (Jinghui et al., 2005).

The selection probability of an individual is based on the equation:

$$ps(a_i) = \frac{f(a_i)}{\sum_{i-1}^{n} f(a_j)}; j = 1, 2, \ldots, n \tag{2.2}$$

where,

- 'n' represents the size of the population.
- $f(a_i)$ represents the fitness value of the individual solution.

Exponential ranking selection

The exponential ranking selection differs from the linear ranking selection technique since the probability selection is formed exponentially. The exponent base is c, where $0 < c < 1$.

$$p_i = \frac{c^{N-i}}{\sum_{j=1}^{N} c^{N-j}}; i \in \{1, \ldots, N\}, \ where \ 0 < c < 1 \tag{2.3}$$

where, c expresses the exponent base.

Crossover (recombination) operators

The crossover operator is based on the exchange of randomized information among parent chromosomes (Singh and Gupta, 2022).

Crossover operators for binary encoding

The most common crossover operators regarding binary genetic algorithms are presented in the following sections.

Single-point crossover

In the single-point crossover a point is randomly selected in the parent chromosomes as a crossover point. The bits on the right of the point are swapped among the parent chromosomes. The generated offsprings, have genetic information from both the parents.

k-point crossover

In the k-point crossover k points are randomly selected in the parent chromosomes as crossover points. The process is equivalent to the two single-point crossovers. The generated offsprings, have genetic information from both the parents.

Uniform crossover

Uniform crossover is implemented by a uniform bit combination of the parents. Each bit is selected from every parent with an equal probability.

Discrete crossover

Discrete crossover operator is similar to the uniform crossover regarding a random number uniformly generated, but the difference is that only one child is generated

Crossover operators for real-coded genetic algorithms

In the following sections, several crossover operators for real-coded genetic algorithms are presented.

Arithmetic crossover

In this crossover operator, an arithmetic operation is utilized to generate a single offspring. The offspring is expressed by the equation:

$$x_i = \alpha a_i + (1-\alpha)b_i, \quad \alpha \in [0,1] \tag{2.4}$$

where,

- a_i, b_i and x_i illustrate the *i*th gene for the parent *A*, the parent *B* and the offspring *X* accordingly.
- *a* (alpha) represents a parameter.

Linear crossover

In the linear crossover, two of the three offsprings are chosen for the next generation. This operator is similar to the arithmetic operation. The equations that express the three offsprings are (Wolpert and Macready, 1997):

$$x_i^1 = -0.5a_i + 1.5b_i \tag{2.5}$$

$$x_i^1 = -0.5a_i + 1.5b_i \tag{2.6}$$

$$x_i^3 = 1.5a_i - 0.5b_i \tag{2.7}$$

Blend crossover

In the blend crossover operator, the offspring is produced by the two parents and for each position, the new gene of the offspring is developed by (Eshelman and Schaffer, 1993):

$$x_i^1 = a_i - \alpha(b_i - a_i) \tag{2.8}$$

$$x_i^2 = b_i + \alpha(b_i - a_i), b_i > a_i \tag{2.9}$$

Simulated binary crossover

The simulated binary crossover (Deb and Agrawal, 1995) is similar to the one-point crossover. The relevant equation for calculating the parameter β is:

$$\beta = \begin{cases} (2u)^{\frac{1}{\eta_c + 1}}, & \text{if } u \leq 0.5 \\ \left(\dfrac{1}{2(1-u)}\right)^{\frac{1}{\eta_c + 1}}, & \text{otherwise} \end{cases} \tag{2.10}$$

where,

- a and b represent the two parents.
- $u \in [0,1)$ expresses a random number.
- η_c represents the distribution index.

The offsprings are calculated by the equations:

$$x^1 = 0.5[(1 + \beta)a + (1 - \beta)b] \tag{2.11}$$

$$x^2 = 0.5[(1 - \beta)a + (1 + \beta)b] \tag{2.12}$$

Three-parent crossover

In the three-parent crossover operator, only one offspring is created by three parent chromosomes.

Mutation operators

In the next sections the most common mutation operators are examined.

Gaussian mutation

The Gaussian mutation operator utilizes a random number generated from a Gaussian distribution. The Gaussian distribution equation is expressed by the equation:

$$f_{G(0,\sigma^2)}(\alpha) = \frac{1}{\sqrt{2\pi\sigma^2}} e^{\frac{\alpha^2}{2\sigma^2}} \qquad (2.13)$$

where,

- σ represents the standard deviation.
- σ^2 represents the variance.
- α represents a random number from a Gaussian distribution in the range [0,1].

Cauchy mutation

The Cauchy mutation operator is based on the Cauchy distribution function to generate the random number. The one-dimensional Cauchy probability density function is expressed by the equation:

$$f(x) = \frac{t^2}{\pi(x^2 + t^2)}, -\infty < x < \infty \qquad (2.14)$$

where t is a positive variable. The Cauchy cumulative distribution function is expressed by the equation:

$$F_t(x) = \frac{1}{2} + \frac{1}{\pi}\arctan\left(\frac{x}{t}\right) \qquad (2.15)$$

The shape of the Cauchy distribution is similar to Gaussian distribution, but the probability is increased in the tails.

Diversity mutation

Diversity mutation was proposed by Deb and Deb (2014). Diversity mutation can improve the exploration properties of any algorithm. An exponential distribution was used:

$$p(i) = te^{-ti}, \text{ with } i \in [0, n-1] \qquad (2.16)$$

the roots of the above probability distribution are given by the equation:

$$te^{-nt} - e^{-t} - t + 1 = 0 \tag{2.17}$$

For n = 15, parameter t = 0.169, for n = 20, t = 0.136. The mutation equation according to the above c is,

$$m(t) = \frac{1}{t} log \left(1 - u\left(1 - e^{-nt}\right)\right) \tag{2.18}$$

Lévy flight mutation

The Lévy flight provides a random walk and the random step length is expressed by the Lévy distribution which is a heavy-tailed:

$$Levy \sim u = t^{-\lambda}, 1 < \lambda \leq 3 \tag{2.19}$$

The equation for Lévy flight mutation is given by (Yang and Deb, 2010):

$$L(x) = 0.01 \times \frac{r_1 \times \sigma}{|r_2|^{1/\beta}} \quad r_1, r_2 \in [0,1]; \beta = 1.5 \tag{2.20}$$

$$\sigma = \left(\frac{\Gamma(1+\beta) \times sin\left(\frac{\pi\beta}{2}\right)}{\Gamma\left(\frac{1+\beta}{2}\right) \times \beta \times 2^{\left(\frac{\beta-1}{2}\right)}} \right)^{1/\beta} \tag{2.21}$$

$$\Gamma(x) = (x-1)! \tag{2.22}$$

where,

- $L(x)$ represents the Lévy distributed random number with variable size x,
- α, β, γ represent random numbers.

Power mutation

Deep and Thakur (2007), proposed the power mutation operator which is based on the power distribution. The distribution function of the power distribution is expressed by the equation:

$$f(x) = px^{p-1}, \quad 0 \leq x \leq 1 \tag{2.23}$$

The density function is expressed by the equation:

$$F(x) = x^p, \quad 0 \leq x \leq 1 \tag{2.24}$$

where p represents the distribution index.

Termination conditions

Fitness limit

The algorithm stops if the best fitness value is less than or equal to the value preset as the fitness limit.

Maximum number of generations

The algorithm execution stops when the iteration number reaches a predefined maximum number.

Maximum stall time

This condition forces the genetic algorithm execution to stop when the best value of the fitness function does not change in the last predefined maximum stall time (E.g., in seconds).

Maximum runtime

The algorithm execution stops when a predefined maximum runtime is reached (E.g., in seconds).

Best fitness value

The algorithm execution stops when the value of the fitness function reaches a predefined best value.

Adaptive genetic algorithm

The crossover probability p_c and the mutation probability p_m are very important factors in genetic algorithm. Finding the optimal values for p_c and p_m is important for improving the genetic algorithm convergence performance (Dey, 2014; Wu et al., 1998; Wu, 2019).

Srinivas and Patnaik (1994), proposed an adaptive strategy, where the crossover probability p_c and the mutation probability p_m are calculated adaptively by the equations:

$$p_c = k_1 \frac{f_{\max} - f_c}{f_{\max} - \overline{f}}, \ k_1 \leq 1.0 \tag{2.25}$$

$$P_m = k_2 \frac{f_{max} - f_i}{f_{max} - \bar{f}}, \; k_2 \leq 1.0 \qquad (2.26)$$

with the constraints,

$$p_c = k_3, \; f_c \leq \bar{f} \qquad (2.27)$$

$$p_m = k_4, \; f_i \leq \bar{f} \qquad (2.28)$$

where,

- k_1, k_2, k_3 and k_4 should be less than 1.0.
- p_c represents the crossover probability.
- p_m represents the mutation probability.
- p_c and p_m are in the range 0.0 to 1.0.
- f_c represents the best fitness values of the selected individuals.
- f_i represents the fitness value of the i_{th} chromosome with probability p_m.
- $\bar{f}$ represents the average fitness value.

Differential evolution

Differential evolution is proposed as a simple and efficient heuristic algorithm in global optimization problems (Storn, 1996; Storn and Price, 1997). Differential evolution algorithm (DE) is a population-based algorithm used in optimization. Differential evolution algorithm is based on an arithmetic operator combined with evolutionary operators: mutation, crossover and selection. Differential evolution can be effective in solving non-convex optimization problems. According to the algorithm, the parameters are initialized, and then mutation, crossover, and selection are performed in every iteration.

The population of differential evolution contains NP D-dimensions vectors for each generation:

$$X_{i,g} = \left(x_{i1,g}, x_{i2,g}, \ldots, x_{iD,g} \right), \; \left(i = 1, \, 2, \ldots, NP \right) \qquad (2.29)$$

where,

- i represents the index of individual
- g represents the number of generations

After initialization, the basic stages of the differential evolution algorithm are: mutation, crossover, and selection (Storn and Price, 1997).

29

Mutation

For each parameter vector $X_{i,g}$ in the population, a mutation vector $V_{i,g}$ is produced according to the following equation:

$$V_{i,g+1} = X_{r_1,g} + F(X_{r_2,g} - X_{r_3,g}); \quad i = 1,2,...,NP \qquad (2.30)$$

where,

- random numbers r_1, r_2 and r_3 are produced with in the range of [1, 2, ...,*NP*], and are mutually different.
- mutation factor or scaling factor $F \in [0, 2]$ is a real constant and controls the differential variation amplification.

Crossover

Differential evolution uses a crossover operation in order to increase the diversity of the population. A trial vector is produced by recombining discretely the parent $X_{i,g}$ and its corresponding mutant vector $V_{i,g}$ using the equations:

$$U_{i,g}, U_{i,g} = \left(u_{i1,g}, u_{i2,g}, ..., u_{iD,g} \right) \qquad (2.31)$$

$$u_{ij,g} = \begin{cases} v_{ij,g}, & if \ (Rand(j) \le CR) \ or \ j = Rnd(i) \\ x_{ij,g}, & otherwise \end{cases} i = 1,2,3,...,NP; \quad j = 1,2,...,D \quad (2.32)$$

where,

- *Rand(j)* represents the *j*-th evaluation of a uniform random number generator between [0, 1].
- *Rnd(i)* represents a randomly chosen value in the range [1, *D*], which is used to ensure that $U_{i,g}$ will select at least one vector from the $V_{i,g}$. in order to help the population evolve.
- *CR* represents the crossover factor within the range [0, 1], which is used to control the recombination.

Selection

The trial vector is compared to the target vector by using greedy selection. The one with the best fitness value will be chosen for the next generation. The selection operation is performed according to the equation:

$$X_{i,g+1} = \begin{cases} U_{i,g} & if \ U_{i,g} \ yields \ better \ fitness \ value \ than \ X_{i,g} \\ X_{i,g}, & otherwise \end{cases} i = 1,2,...,NP \quad (2.33)$$

Chaotic differential evolution

Several researchers have proposed Chaotic differential evolution algorithms to enhance the optimization ability of the original differential evolution algorithm. Lu et al. (2011). The performance of differential evolution is affected by the value of mutation factor F and the crossover factor (CR). The use of chaotic sequences to adapt the parameter value settings can enhance differential evolution and enhance the convergence capability of the algorithm.

Lu et al. (2011), proposed Tent equation to be used to adjust the values of F and CR in a dynamical way. The Tent equation is expressed by the equation:

$$cx_i^{k+1} = \begin{cases} 2cx_i^k & 0 \le cx_i^k \le 0.5 \\ 2(1-cx_i^k) & 0.5 < cx_i^k \le 1 \end{cases} i = 1, 2, \ldots, D; \quad k = 1, 2, \ldots, k_{max} \quad (2.34)$$

where,

- k represents the number of the iteration
- k_{max} represents the maximum iteration of the chaotic sequences
- cx_i^k represents the i-th chaotic parameter with values within the range $[0, 1]$.

The conditions are: the initial $cx_i^0 \in (0, 1)$ and $cx_i^0 \notin \{1/4, 1/2, 2/3, 3/4\}$.

Lu et al. (2011), proposed the following equation for mutation factor F in order to improve the differential evolution efficiency:

$$F^0 \in (0,1), F^0 \notin \{1/4, 1/2, 2/3, 3/4\} \quad (2.35)$$

$$F^{g+1} = \begin{cases} 2F^g & 0 < F^g < 0.5 \\ 2(1-F^g) & 0.5 < F^g < 1 \end{cases} g = 1, 2, \ldots, g_{max} \quad (2.36)$$

Similarly, Lu et al. (2011), proposed the following equations for crossover factor CR in order to improve the differential evolution efficiency:

$$CR^0 \in (0,1), CR^0 \notin \{1/4, 1/2, 2/3, 3/4\} \quad (2.37)$$

$$CR^{g+1} = \begin{cases} 2CR^g & 0 < CR^g < 0.5 \\ 2(1-CR^g) & 0.5 < CR^g < 1 \end{cases} g = 1, 2, \ldots, g_{max} \quad (2.38)$$

31

Coelho and Mariani (2006), proposed the utilization of the logistic map as a chaotic map for dynamically adapting the mutation factor F:

$$y(t) = \mu \cdot y(t-1) \cdot \left[1 - y(t-1)\right] \tag{2.39}$$

dos Santos et al. (2012), proposed a chaotic differential evolution algorithm based on Ikeda map which generates chaotic sequences in order to tune the CR and the F parameters. The Ikeda map is expressed by the equations:

$$X(k+1) = 1 + \lambda \cdot [X(k)\cos(c(k)) - Y(k)\sin(c(k))] \tag{2.40}$$

$$Y(k+1) = \lambda \cdot [X(k)\sin(c(k)) + Y(k)\cos(c(k))] \tag{2.41}$$

where,

- λ is a parameter with the value 0.9
- $c(k)$ is expressed by the equation:

$$c(k) = 0.4 - \frac{6}{1 + X(k)^2 + Y(k)^2} \tag{2.42}$$

Differential evolution example

Ackley function approximation

Mathematical equation

The equation that represents the Ackley function is the following.

$$f(x_1,...,x_n) = -a \cdot exp(-b\sqrt{\frac{1}{n}\sum_{i=1}^{n}x_i^2}) - exp(\frac{1}{n}\sum_{i=1}^{n}cos(cx_i)) + a + exp(1) \tag{2.43}$$

where the a, b and c are constant values. For the experiments we set: a = 20, b = 0.2 and c = 2π. The Ackley function is continuous, non-convex, multimodal and differentiable.

Input search space

The function is evaluated with $x_i \in$ [−20, 20], and i = 1, 2, . . . , n.

Differential evolution parameters

Maximum Number of Iterations: 2000

Population Size: 50

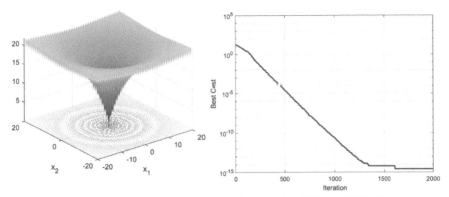

Figure 2.1. Ackley function approximation by using Differential Evolution.

Decision Variables Minimum Bound: −20

Decision Variables Maximum Bound: 20

Crossover Probability: 0.2

Lower Bound of Scaling Factor: 0.2

Upper Bound of Scaling Factor: 0.8

The optimization results illustrated in Figure 2.1 are very good. The function minimization stops improving at about 2000 iterations.

References

Alavi, M. and Henderson, J. C. (1981). An evolutionary strategy for implementing a decision support system. *Management Science*, 27(11): 1309–1323.

Back, T. (1996). *Evolutionary Algorithms in Theory and Practice: Evolution Strategies, Evolutionary Programming, Genetic Algorithms*. Oxford University Press.

Baker, B. M. and Ayechew, M. A. (2003). A genetic algorithm for the vehicle routing problem. *Computers & Operations Research*, 30(5): 787–800. doi:https://doi.org/10.1016/S0305-0548(02)00051-5.

Baker, J. E. (1985, July). Adaptive selection methods for genetic algorithms. *In: Proceedings of an International Conference on Genetic Algorithms and their Applications* (Vol. 1).

Berger, J. and Barkaoui, M. (2004). A parallel hybrid genetic algorithm for the vehicle routing problem with time windows. *Computers & Operations Research*, 31(12): 2037–2053.

Coelho, L. S. and Mariani, V. C. (2006). Combining of chaotic differential evolution and quadratic programming for economic dispatch optimization with valve-point effect. *IEEE Transactions on Power Systems*, 21(2): 989–996. doi:10.1109/TPWRS.2006.873410.

Deb, K. and Agrawal, R. B. (1995). Simulated binary crossover for continuous search space. *Complex Systems*, 9(2): 115–148.

Deb, K. and Deb, D. (2014). Analysing mutation schemes for real-parameter genetic algorithms. *International Journal of Artificial Intelligence and Soft Computing*, 4(1): 1–28.

Deep, K. and Thakur, M. (2007). A new mutation operator for real coded genetic algorithms. *Applied Mathematics and Computation*, 193(1): 211–230.

Dey, D. K. (2014). Mathematical study of adaptive genetic algorithm (AGA) with mutation and crossover probabilities. *Compusoft*, 3(5): 765.

dos Santos, G. S., Luvizotto, L. G. J., Mariani, V. C. and Coelho, L. d. S. (2012). Least squares support vector machines with tuning based on chaotic differential evolution approach applied to the identification of a thermal process. *Expert Systems with Applications*, 39(5): 4805–4812. doi:https://doi.org/10.1016/j.eswa.2011.09.137.

Eshelman, L. J. and Schaffer, J. D. (1993). Real-coded genetic algorithms and interval-schemata. pp. 187–202. *In: Foundations of Genetic Algorithms* (Vol. 2). Elsevier.

Ferentinos, K. P. and Tsiligiridis, T. A. (2007). Adaptive design optimization of wireless sensor networks using genetic algorithms. Elsevier, *Computer Networks*, 51(4): 1031–1051.

Goldberg, D. E. and Deb, K. (1991). A comparative analysis of selection schemes used in genetic algorithms. pp. 69–93. *In: Foundations of Genetic Algorithms* (Vol. 1). Elsevier.

Holland, J. H. (1975). *Adaptation in Natural and Artificial Systems. An Introductory Analysis with Applications to Biology, Control and Artificial Intelligence*. Ann Arbor: University of Michigan Press.

Lamini, C., Benhlima, S. and Elbekri, A. (2018). Genetic algorithm based approach for autonomous mobile robot path planning. *Procedia Computer Science*, 127: 180–189.

Lu, Y., Zhou, J., Qin, H., Wang, Y., Zhang, Y. et al. (2011). Chaotic differential evolution methods for dynamic economic dispatch with valve-point effects. *Engineering Applications of Artificial Intelligence*, 24(2): 378–387. doi:https://doi.org/10.1016/j.engappai.2010.10.014.

May, A. T., Jariyavajee, C. and Polvichai, J. (2021, December). An improved genetic algorithm for vehicle routing problem with hard time windows. pp. 1–6. In 2021 International Conference on Electrical, Computer and Energy Technologies (ICECET). IEEE.

Negnevitsky, M. (2005). *Artificial Intelligence: a Guide to Intelligent Systems*. Pearson Education.

Rechenberg, I. (1965). Cybernetic solution path of an experimental problem. *Royal Aircraft Establishment Library Translation 1122*.

Rodriguez, A., Falcarin, P. and Ordonez, A. (2015, November). Energy optimization in wireless sensor networks based on genetic algorithms. pp. 470–474. In 2015 SAI Intelligent Systems Conference (IntelliSys). IEEE.

Sahay, K. B., Sonkar, A. and Kumar, A. (2018, October). Economic load dispatch using genetic algorithm optimization technique. pp. 1–5. In 2018 International Conference and Utility Exhibition on Green Energy for Sustainable Development (ICUE). IEEE.

Schwefel, H. P. (1981). *Numerical Optimization of Computer Models*. John Wiley & Sons, Inc.

Schwefel, H. P. (1995). *Evolution and Optimum Seeking*. John Wiley and Sons. New York.

Singh, G. and Gupta, N. (2022). A study of crossover operators in genetic algorithms. pp. 17–32. *In*: *Frontiers in Nature-Inspired Industrial Optimization*. Springer, Singapore

Srinivas, M. and Patnaik, L. M. (1994). Adaptive probabilities of crossover and mutation in genetic algorithms. *IEEE Transactions on Systems, Man, and Cybernetics*, 24(4): 656–667.

Storn, R. (1996, June). On the usage of differential evolution for function optimization. pp. 519–523. *In*: *Proceedings of North American Fuzzy Information Processing*. IEEE.

Storn, R. and Price, K. (1997). Differential evolution—a simple and efficient heuristic for global optimization over continuous spaces. *Journal of Global Optimization*, 11(4): 341–359.

Wolpert, D. H. and Macready, W. G. (1997). No free lunch theorems for optimization. *IEEE Transactions on Evolutionary Computation*, 1(1): 67–82.

Wu, Q. H., Cao, Y. J. and Wen, J. Y. (1998). Optimal reactive power dispatch using an adaptive genetic algorithm. *International Journal of Electrical Power & Energy Systems*, 20(8): 563–569. doi:https://doi.org/10.1016/S0142-0615(98)00016-7.

Wu, X. (2019, July). Research on the influence of crossover probability and mutation probability in GA-SVM Model. pp. 1–7. *In*: *2019 IEEE 9th International Conference on Electronics Information and Emergency Communication (ICEIEC)*. IEEE.

Yang, X. S. and Deb, S. (2010). Engineering optimisation by cuckoo search. *International Journal of Mathematical Modelling and Numerical Optimisation*, 1(4): 330–343.

CHAPTER 3

Swarm Intelligence and Particle Swarm Optimization

<><><><><><><><><><><><><><><><><><><><><><><><><><><><><><><><><><><><><><><><><><><><><><><><><>

This chapter is devoted to the theoretical foundation of particle swarm optimization, the latest PSO algorithm advances, neural network optimization methods and illustrative examples. The standard bio-inspired particle swarm algorithm is presented and some important variations of the PSO algorithm are explained and analyzed such as: Quantum PSO algorithm with delta potential well and harmonic oscillator, Chaotic PSO, mutation-based PSO and multi-objective PSO. Advanced aspects and variants of particle swarm optimization are discussed, such as: swarm topologies, boundary approaches, swarm initialization techniques, stopping criteria and swarm mutation operators, bringing to the reader a complete presentation of the PSO algorithm and its latest variations. Illustrative application examples are presented regarding unimodal, multimodal, convex and non-convex function approximation.

Particle swarm optimization algorithm

The Particle Swarm Optimization (PSO) was initially proposed by Eberhart and Kennedy in 1995 (Eberhart and Kennedy, 1995). The PSO algorithm was inspired by studying the swarm behavior of animals in nature. The natural social intelligence techniques were suitably transformed in order to develop a computational optimization algorithm. Particle Swarm Optimization is an evolutionary algorithm with a stochastic nature where the swarm particles are traversing a predefined search space to evaluate the fitness function. Each swarm particle is characterized by its position and velocity.

Each particle velocity is updated according to the equation (Eberhart, 1995; Eberhart et al., 2001):

$$v_i(t + 1) = w\, v_i(t) + c_1\, r_1[\hat{x}ibest_i(t) - x_i(t)] + c_2\, r_2[gbest(t) - x_i(t)] \quad (3.1)$$

where,

t represents the time

$v_i(t)$ illustrates the particle velocity at time t

$v_i(t + 1)$ illustrates the particle velocity at time t+1

$x_i(t)$ illustrates the particle position at time t

$\hat{x}ibest_i(t)$ illustrates the particle individual best objective value solution as of time t

r_1, r_2 represents random numbers uniformly distributed in the interval (0,1)

$gbest(t)$ represents the swarm's global best objective value solution as of time t

c_1 illustrates the coefficient parameter of personal learning

c_2 illustrates the coefficient parameter of global learning

w illustrates the inertia weight coefficient parameter

Also, the particle position in the next time step $t+1$ is calculated by using the following equation:

$$x_i(t + 1) = x_i(t) + v_i(t + 1) \quad\quad\quad (3.2)$$

The standard particle swarm optimization algorithm consists of the following basic steps:

1. Initialization process:
   a. Initialize the particle swarm population
   b. Initialize the particle topology and positions
   c. Initialize the particle velocities
   d. Evaluate each particle's fitness function and initialize global best solution
2. Execute the following steps until the stopping criterion of the algorithm is met:
   a. Set the coefficient numbers: r_1, r_2 randomly in the space (0,1)
   b. Update the particle velocity according to eq. (1)
   c. Update the particle position according to eq. (2)

    d. Evaluate each particle's fitness function

    e. If the particle's personal best is better, update the current personal best

    f. If the swarm global best is better, update the current global best

3. Stop the algorithm process and show the global best solution

Hyper-parameters

Acceleration coefficients

The c_1 and c_2 coefficients of personal and global learning, are also called as acceleration coefficients. These coefficients are combined with the random parameters r_1 and r_2 in eq. (1) to control the influence of the personal and the social factors on the particle swarm velocity vector. The constant c_1 expresses the amount of influence of the personal particle cognition and the constant c_2 expresses the amount of influence of the social neighborhood affection on the swarm particle. Also, the acceleration coefficients are combined with the random parameters r_1 and r_2, and control the cognitive and social influence on the particle velocity.

In the standard PSO algorithm, constant values of acceleration coefficients were proposed during the whole optimization process in the search space. Later several researchers have proposed adaptive acceleration coefficients. For example, Chen et al. (2014), have proposed an adaptive acceleration coefficient in order to improve the algorithm search capability. The c_1 coefficient decreases from the c_{1i} initial value to the final value c_{1f}, and the c_2 coefficient increases from the c_{2i} initial value to c_{2f} final value by applying the following equations:

$$c_1 = (c_{1f} - c_{1i})\frac{t}{t_{max}} + c_{1i} \qquad (3.3)$$

$$c_2 = (c_{2f} - c_{2i})\frac{t}{t_{max}} + c_{2i} \qquad (3.4)$$

where t represents the current iteration, and t_{max} represents the maximum number of iterations.

Inertia weight

The inertia weight is used to control the speed of the flying swarm particles. When the swarm particle speed is very big the optimal fitness solution can be easily skipped although the swarm convergence is very quick. On the other hand, when the particle speed is very slow, the swarm convergence is also slow and can be trapped easily in a local minimum.

Van den Bergh and Engelbrecht (2006), have proved that all inertia weight values which are larger than a critical value, guarantee convergent particle swarm trajectories. This critical value is calculated by the equation:

$$w > \frac{1}{2}(c_1 + c_2) - 1 \tag{3.5}$$

Initially, a static value of inertia weight was proposed during the entire optimization in the search space for all swarm particles. Later several researchers proposed adaptive or dynamically changing inertia weight. Several variations of inertia weight were proposed to improve the search ability and convergence of the PSO algorithm (Chen et al., 2014; Shi and Eberhart, 1998; Tian 2021). Shi and Eberhart (1998), proposed another type of inertia weight, called time-varying inertia weight. The proposed inertia weight decreases through time and the experimental results have shown a significant improvement in the performance of the standard PSO.

Tian (2021), has proposed an adaptive inertia weight expressed by the following equation:

$$\omega = \begin{cases} \omega_{min} - (\omega_{max} - \omega_{min}) \exp\left[\left(-n(f - f_{min})\right)\right], f \le f_{avg} \\ \omega_{max}, f > f_{avg} \end{cases} \tag{3.6}$$

where,

- n is the iteration number
- f represents the current fitness function value
- f_{avg} represents the average fitness function value of all particles.

The modified inertia weight has the advantage of better adaptation during the optimization process since its value changes in every iteration compared to the constant weight of the standard PSO algorithm.

Chen et al. (2014), have proposed another adaptive inertia weight where the weight value is reducing during the iterations. This way the PSO algorithm can investigate specific areas in the local search space. The inertia weight wt is calculated by using the following equation:

$$wt = wt_{min} + (wt_{max} - wt_{min}) \frac{(t_{max} - t)}{t_{max}} \tag{3.7}$$

where,

wt_{max} is a predefined maximum inertia weight value

wt_{min} represents a predefined minimum inertia weight value

t is the current iteration of the algorithm

t_{max} is the maximum number of iterations.

Stopping criteria

In this section, the stopping criteria of the PSO algorithm are examined. The stopping criteria are specific conditions which are set in order to terminate the PSO algorithm process when the swarm convergence is met. The stopping criteria are as follows.

Maximum generation number

The algorithm execution stops when the iteration number reaches a predefined maximum number.

Maximum stall time

This condition forces PSO algorithm execution to stop when the best value of the fitness function does not change in the last predefined maximum stall time (E.g., in seconds).

Maximum runtime

The algorithm execution stops when a predefined maximum runtime is reached (E.g., in seconds).

Best fitness value

The algorithm execution stops when the value of the fitness function reaches a predefined best value.

Population convergence

This criterion stops the algorithm execution when the difference between the maximum and minimum values of all the particles in the population is less than the limit tolerance value.

Fitness convergence

This criterion stops the algorithm execution when the difference between the maximum and minimum fitness values is less than the limit tolerance value.

Figure 3.1, illustrates an example of how the swarm particles traverse the search space in order to find the best value of an objective function.

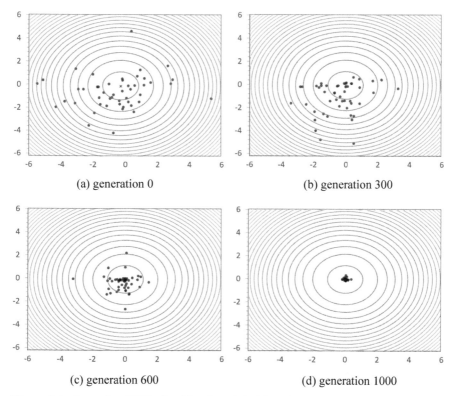

Figure 3.1. Example of PSO algorithm for minimizing the rastrigin function. The swarm particles traverse the search space in order to minimize the objective function. The population size was set to 200.

Swarm topologies

In this section, the most important PSO topologies are discussed. The swarm topology is a very important factor which affects the swarm convergence. The swarm topologies the are discussed are static topologies which do not alter during the optimization process such as: Global or Fully connected topology, local or ring, Von Neumann, Star, Mesh, Random,

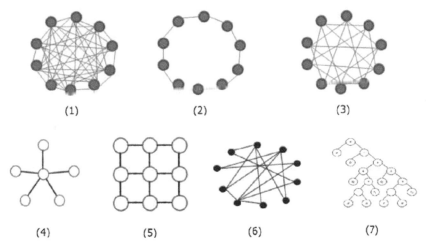

Figure 3.2. Illustration of swarm topologies: (1) Global or Fully connected, (2) Ring, (3) Von Neumann, (4) Star, (5) Mesh, (6) Random, (7) Tree or Hierarchical.

Tree or Hierarchical topology and also dynamic topologies which change during the convergence process, such as: multi-ring and clan topologies.

Global or fully connected

In a global topology, the swarm individuals are fully connected. Each particle is connected and exchange information with all the other particles. Consequently, in this structure, faster convergence is achieved compared to other topologies since particles in the swarm can communicate with all the other particles faster to find the global best solution, but the convergence process can be trapped in local best.

Local or ring topology

The ring swarm topology corresponds to a local best architecture of PSO. In this structure, every individual particle interacts with its N adjacent neighbor particles. This means that each individual particle searches for the best objective value locally, only in its neighborhood. When $N = 2$ there are two adjacent neighbor particles as illustrated in the figure for ring topology. In this topology, a local neighborhood could converge in different local optima as opposed to another local neighborhood which could be still searching for the best local solution. The advantage of this structure is that different local search spaces of the global search space can be explored simultaneously (Kennedy and Mendes, 2002).

43

Von Neumann

In Von Neumann topology, each swarm particle is communicating with four other individual particles, the two prior and the two following particles. Some studies, have shown that this topology can produce better results compared to others (Bratton and Kennedy, 2007; Kennedy and Mendes, 2002).

Star topology

In star topology, the communication of the individual particles is accomplished through only one particle. This central particle is affected by all the other swarm particles. The star structure, leads to particle isolation since all the information about the best objective value is exchanged through the central swarm particle. The central particle fathers and compares the local bests of each swarm particle and adjusts its new best position.

Mesh topology

In this topology, one swarm particle is connected to four neighbouring particles. The swarm particles on the edges are connected with two adjacent swarm particles. This means that there are overlapping particle neighbors which create redundancy during the convergence process. The mesh architecture of the swarm particles remains during the algorithm execution process.

Random topology

In random topology, the swarm individuals are connected with a random number of swarm particles other than them. Each particle is connected and exchanges information only with the connected particles. This structure does not change in all the convergence processes.

Tree or hierarchical topology

In the hierarchical or tree topology there are hierarchic levels of swarm particles. Also, there is a central particle which functions as a root in the top level and it is connected with the particles of the next lower level in the hierarchy. The tree topology is similar to the binary tree architecture. The root particle is selected randomly in the swarm and searches for the best solution obtained by the next lower level in the hierarchy.

Dynamic or adaptive topologies

All the previous topologies are static since they cannot be changed in all the optimization processes. Another category of topologies is based on a dynamic mode. More recently, dynamics topologies have been proposed. These neighborhood topologies vary during the optimization process. Some of them are: Multi-Ring and Clan topology. In Multi-Ring topology (Bastos-Filho et al., 2008), which is constructed by several layers placed one over the other with each having a ring topology. In Clan topology (Carvalho et al., 2009), where the particle swarm is separated into sub-swarms named as 'clans' and every clan has a fully-connected architecture. Also, there are other adaptive topologies such as: adaptive population-based PSO (Almasi and Khooban, 2018), complementary Cyber Swarm (Yin et al., 2011), stochastic star PSO (Miranda et al., 2008).

Boundary handling approaches

In this section, several boundary handling approaches are discussed which affect the performance of the PSO algorithm. These techniques are very important since they can restrict the swarm particle movement in the predefined search space (Helwig et al., 2013; Mostaghim et al., 2006; Padhye et al., 2013). Therefore, these methods can help the algorithm achieve convergence faster to a global solution.

Hyperbolic method

In this technique, the swarm particle cannot leave the feasible space (Clerc, 2006). This is achieved by normalizing the velocity of the swarm particle according to the equations:

$$v_{i,j}^{t+1} = \begin{cases} \dfrac{v_{i,j}^{t+1}}{1+|v_{i,j}^{t+1}/(X_{max}-x_{i,j}^{t})|} & \text{when } v_{ij}^{t+1} > 0 \\[3ex] \dfrac{v_{i,j}^{t+1}}{1+|v_{i,j}^{t+1}/(x_{i,j}^{t}-X_{min})|} & \text{when } v_{ij}^{t+1} \leq 0 \end{cases} \tag{3.8}$$

where, the X_{max} and X_{min} parameters represent the upper and lower boundaries in the jth dimension. The new normalized particle velocity secures the new swarm particle position such that it cannot be out of bounds. This method restricts the velocity of the swarm particle in the feasible space and converges faster.

Infinity or invisible wall

In this method the only modification is that the personal best and the global best particle positions should be in the search space (Bratton and

Kennedy, 2007; Robinson and Rahmat-Samii, 2004). This means that, if the particles leave the search space, their local and global best attractors will be within search bounds. The infinity technique does not modify the particle velocity or position, and as a sequence the algorithm is not affected by the boundary handling method.

Nearest or boundary or absorb

In this technique, the swarm particle that violates the boundary is moved back to the boundary of the violated parameter. The result when applying this method, is that the swarm particles are moved near the problem solutions that are close to the search space boundary.

Random

In this method if a particle is out of bounds then the position is reinitialized uniformly in a valid position within the problem search space. This way the diversity of swarm particle solutions is increased by randomly placing the particles in positions that would have been unlikely to be placed otherwise.

Random-half

In this method if a particle is out of bounds, then the position is reinitialized uniformly to a valid position within half the search space nearest to the violated point.

Periodic

This method provides an infinite space, which includes periodic copies of the original search space (Zhang et al., 2004). Each particle which violates the boundaries is placed to a location composed of copies of the original search space. The Periodic technique handles the particle variables and allows the particle to enter the search space again opposite to the place where it has left. This method has the advantage of eliminating particle swarm disorganization compared to other conventional methods such as mutation-based boundary handling methods. The performance of particle swarm optimization with the Periodic boundary handling method has shown improved optimization results compared to Random and Boundary methods, especially when the local optimums are located close to the boundary.

Exponential

This method provides a correction of the particle position parameter in every dimension by placing the swarm particle to a location among the previous particle positions and the violated boundary position (Alvarez-Benítez et al., 2005). The new position is defined by implementing a truncated exponential distribution, in order to have increased probability of being placed in a position closer to the boundary. The exponential method preserves search data regarding the best solutions close to the boundaries compared to the random and random-half methods.

Mutation

This method includes a mutation that has been utilized so that the violated particle parameter can be perturbed in order to be placed back into the problem search region.

Reflect methods

In the reflect technique, the positions of the violated particle parameters are reflected in the closer boundary into the search space. In the reflected position the new velocity can be set with several methods in the violated variables:

- Unmodified velocity
- Velocity set to zero
- Velocity adjusted according to the equation: $V_{new} = V_{new,bounded} - V_{old,unbounded}$

Random damping

Huang and Mohan (2005), have proposed a hybrid boundary condition in particle swarm optimization. If a swarm particle violates the boundary, then the particle velocity is partially absorbed and partially reflected. This way the particle is forced to return into valid space with decreased velocity. The fraction of the reflected and absorbed velocity is random.

PSO with mutation

In the standard PSO algorithm there are no evolution parameters like mutation and crossover. Although, several researchers in the literature have proposed PSO algorithms with mutation operators (Assarzadeh and

Naghsh-Nilchi, 2015; Lin and Zhang, 2017; Stacey et al., 2003) they can contribute to overcome early convergence into local minimums that could appear in the standard particle swarm optimization algorithm. In this section, a review of researches that have improved the standard PSO algorithm by adding a mutation strategy have been discussed.

Gaussian mutation PSO

Higashi and Iba (2003), have applied a mutation operator which alters the value of a particle swarm dimension by using a random number from a Gaussian distribution. The equation which changes the particle position dimensions, x_{id}, is expressed by:

$$mutate(x_{id}) = x_{id}(1 + gaussian(\sigma))$$

where,

- σ represents the standard deviation
- gaussian(σ) represents the function which draws a random number from a Gaussian distribution

Cauchy mutation PSO

Stacey et al. (2003), have proposed a similar mutation operator by implementing the Cauchy probability distribution. The Cauchy mutation is applied on the particle with the global best value in every generation. The one-dimensional Cauchy probability density function is expressed by the equation:

$$f(x) = \frac{t^2}{\pi(x^2 + t^2)}, -\infty < x < \infty \tag{3.9}$$

where t is a positive parameter. The Cauchy cumulative distribution function is expressed by the equation:

$$F_t(x) = \frac{1}{2} + \frac{1}{\pi} \arctan\left(\frac{x}{t}\right) \tag{3.10}$$

The shape of the Cauchy distribution is similar to the Gaussian distribution, but the probability is more in the tails. The x_{id} particle position dimensions, are updated by the equation:

$$mutate(x_{id}) = x_{id} + cauchy(a) \tag{3.11}$$

where a represents a parameter similar to the σ parameter of Gaussian distribution that affects the shape of the distribution.

Michalewicz non-uniform mutation

Esquivel and Coello (2003), have studied a non-uniform mutation operator for PSO developed by Michalewicz to be used in genetic algorithms (Michalewicz, 1998; Dasgupta and Michalewicz, 2013). The Michalewicz's non-uniform mutation operator uses random numbers non-uniformly to mutate values and it is useful when the PSO algorithm is trapped in local optima and when highly multimodal functions are being optimized. In the Michalewicz's non-uniform operator, the particle position dimension, x_{id}, is mutated by using the equation:

$$\text{mutate}(x_{id}) = \begin{cases} x_{id} + \text{delta}(it, U - x_{id}) : rb = 1 \\ x_{id} + \text{delta}(it, U - x_{id}) : rb = 1 \end{cases} \qquad (3.12)$$

where,

- *it* illustrates the iteration number
- *U* represents the upper bound
- *L* represents the lower bound
- *rb* the generated random bit
- delta (t,y) is calculated by the equation:

$$\text{delta}(t, y) = y(1 - r^{(1-t/T)^b}) \qquad (3.13)$$

where *r* represents a randomly generated number in the range [0,1], *T* the maximum iteration number of the algorithm, and *b* represents a parameter usually set to the value 5 (Michalewicz, 1998).

Chaotic PSO with Michalewicz mutation

In the swarm optimization algorithms, which are modified based on the chaos theory, the basic application technique is to use chaotic parameters instead of the usual random variables of the PSO algorithm. Assarzadeh and Naghsh-Nilchi (2015), have proposed the application of a mutation method to a chaotic particle swarm optimization for classification problems. The study proposes the implementation of logistic maps as chaotic sequences by combining also the Michalewicz's nonuniform mutation. In this research, the random parameters in PSO are replaced by the chaotic sequences generated by the logistic map. The random parameters are given by the logistic map chaotic equation:

$$Cr_{(t+1)} = kCr_{(t)}(1 - Cr_{(t)}) \qquad (3.14)$$

where n = 0,1,2,3... and k = 4. The Cr(t) parameter is controlled by the k parameter of the chaotic logistic map. Considering the above assumptions the particle velocity update equation of the chaotic particle swarm optimization can be expressed as:

$$v_i(t+1) = w\,v_i(t) + c_1 C_r[Pbest(t) - x_i(t)] + c_2(1 - C_r) + c_2(1 - C_r)[Gbest(t) - x_i(t)]$$

$$(3.15)$$

Final results of the research have shown an improved performance of Chaotic PSO with Michalewicz mutation in high dimensional spaces as a global search optimization algorithm.

Random mutation PSO

In this kind of mutation, the operator initializes the particle swarm position with a randomly uniformly generated value in the range (Andrews, 2006).

Constant mutation PSO

In this type of mutation, the mutation rate remains the same during the optimization iterations.

Stagnant mutation

This kind of mutation is applied when the global best of the particle population remains the same for a fixed number of iterations.

Quantum PSO

Quantum PSO (QPSO) was introduced in 2004 (Sun et al., 2004). Quantum-behaved PSO is a very interesting variation of the standard PSO algorithm. The QPSO algorithm incorporates quantum mechanics in the PSO algorithm. In classical mechanics, the swarm particle is defined by the vectors regarding the position x_i and the velocity v_i of the particle in the swarm. In quantum mechanics, the position x_i and the velocity v_i of the particle cannot be defined simultaneously because of the uncertainty principle (Sun et al., 2011). Consequently, in a quantum system the PSO particles will have a quantum behavior, which means that the particles will be defined by their state and not by their position and velocity. The state of the swarm particle is expressed by the using the wavefunction $\psi(x,t)$ of quantum mechanics. In the QPSO algorithm the swarm is a quantum system and every particle is identified by a quantum state. In the following sections, two quantum PSO algorithms are discussed: Delta well quantum PSO and Harmonic quantum PSO.

Delta well quantum PSO

In the delta well quantum PSO principle (Sun et al., 2011) the swarm particle is moving to a position p in a one-dimensional delta potential field. The delta potential well field is expressed by the equation:

$$V(x) = \begin{cases} \infty & \text{if } x \text{ is } 0 \\ 0 & \text{if } x \text{ is} \neq 0 \end{cases} \tag{3.16}$$

By solving the Schrödinger equation in a one-dimensional delta potential well, the probability density function can be found.

$$D(x_{i,j}(t+1)) = \frac{1}{L_{i,j}(t)} e^{-2|p_{i,j}(t)-x_{i,j}(t+1)|/L_{i,j}(t)} \tag{3.17}$$

$$F(x_{i,j}(t+1)) = e^{-2|p_{i,j}(t)-x_{i,j}(t+1)|/L_{i,j}(t)} \tag{3.18}$$

where,
- D represents the normalized probability density function
- $L(t)$ represents the characterization potential well length

Then the Monte Carlo method is implemented to find the particle position according to the equation principle (Sun et al., 2011):

$$\begin{cases} x_i(t+1) = p_i + \beta \cdot | Pbest_i - x_i(t)| \cdot \ln(1/u) \\ x_i(t+1) = p_i - \beta \cdot | Pbest_i - x_i(t)| \cdot \ln(1/u) \end{cases} \tag{3.19}$$

- where β is the contraction-expansion parameter (Sun et al., 2011);
- u is a random number in the uniformly distributed space [0,1]
- $Pbest_i$ is mean of the pbest swarm positions
- p_i is the best position of the ith swarm particle

The global mean best particle (*Pbest*) of the swarm is expressed by the mean of all the particle best positions:

$$Pbest_i = \frac{1}{N} \sum_{i=1}^{N} p_{k,i}(t) \tag{3.20}$$

where k illustrates the best particle index in the swarm of N particles.

The Delta well quantum PSO algorithm includes the following steps:

Step 1. Particle swarm initialization of the population and random particle positions uniformly distributed in the search space.

Step 2. Particle fitness function evaluation.

Step 3. Comparison of personal best value of the fitness function. If the current fitness function value is better, then set the personal best value equal to the current fitness function value in the search space.

Step 4. Compare the current global best solution with the previous global best solution. If the current value is better then set it as the global best.

Step 5. Update the global best point by estimating the Pbest according to relevant equation.

Step 6. Update the particles' positions.

Step 7. Repeat the process from step 2 until a stop criterion is achieved, e.g., the maximum number of iterations.

Harmonic quantum PSO

Several other potential fields were proposed. In the harmonic quantum PSO (Sabat et al., 2010) the harmonic oscillator is utilized as a potential field. In the Harmonic potential distribution, the potential well energy of the swarm particle in a one-dimensional harmonic potential well is expressed by the equation:

$$V(x) = \frac{1}{2}kx^2 \qquad (3.21)$$

where, k represents the harmonic well depth.

The D probability density function is given by the equation:

$$D(x) = |\psi(x)|^2 = L\exp(-0.5\alpha^2 x^2) \qquad (3.22)$$

$$L = \frac{\alpha}{\sqrt{\pi}} \qquad (3.23)$$

where L is the characteristic well length. The new quantum particle position can be found by using the equation (Sabat et al., 2010):

$$x_{i,j}(t+1) = p_{i,j}(t) \pm \alpha \cdot \frac{1}{0.4769g} |x_{i,j}(t) - mpbest_j(t)| \sqrt{\ln\frac{1}{u}} \qquad (3.24)$$

where,

- mpbest is the mean swarm best of particles
- g is a parameter set to 2.0.

Similarly, with delta well quantum PSO, the implementation involves the Monte Carlo method. The Harmonic QPSO algorithm implementation is similar to the delta well quantum PSO.

Multi-objective PSO

A multi-objective optimization problem with several conflicting objectives can be defined as follows:

$$\text{minimize } F(\mathbf{x}) = \left(f_1(\mathbf{x}), f_2(\mathbf{x}), \ldots, f_m(\mathbf{x}) \right)^{\mathrm{T}} \qquad (3.25)$$

$$\text{subject to } \mathbf{x} \in \Omega \qquad (3.26)$$

where Ω represents the feasible set with:

$$\Omega \subseteq \mathbb{R}^n \neq \varnothing \qquad (3.27)$$

and

$$f_i : \mathbb{R}^n \to \mathbb{R} \qquad (3.28)$$

represents the m objective functions for $i = 1, 2, \ldots, m$, and $m \geq 2$. The objective space is the image of the feasible set:

$$\mathcal{Y} = \left\{ F(x) \in \mathbb{R}^m : x \in \Omega \right\} \qquad (3.29)$$

The sets R^n and R^m represent the *decision space* and the *objective space* respectively.

In multi-objective optimization problems, the dominance defines the degree of the goodness of a solution. The concept of dominance is explained generally as: A point x dominates another point y in the objective function vector f when:

$$f_i(x) \leq f_i(y) \text{ for all } i \qquad (3.30)$$

$$f_j(x) < f_j(y) \text{ for some } j \qquad (3.31)$$

A nondominated set of solutions (points) among a set of solutions P represents the set of solutions Q in set of solutions P that are not dominated by any other solution in P.

A particle solution $x' \in X$ represents a non-dominated solution if no other particle solution $x \in X$ exists such that:

$$f_i(x) \leqslant f_i(x') \text{ for all } i=1,\dots,m \qquad (3.32)$$

and at least one index i exists such that:

$$f_i(x) < f_i(x') \qquad (3.33)$$

In a given set of swarm particle solutions, the non-dominated solution set represents the set of all the swarm particle solutions which are not dominated by any other solution of the set. The non-dominated set of the whole feasible decision space represents the Pareto optimal set of the swarm particle solutions. The decision boundary determined by the set of all points (solutions) mapped from the Pareto optimal set represents the Pareto optimal front (Deb, 2011).

In multi-objective problems, when a solution is not dominated by another solution it is a Pareto optimal solution. The discovered non-dominated particle solutions are stored in the external global memory or repository. The steps of the multi-objective PSO algorithm (Coello and Lechuga, 2002; Coello et al., 2004) are described as follows:

1. Initialize the particle population positions *POPi*, where i = [1, 2,...,N] and N represents the particle population size.
2. Initialize velocity *VELi* of each swarm particle
3. Evaluate each swarm particle in the *POPi*.
4. Save the positions of the particles which express the non-dominated solutions in the repository (*REP*).
5. Generate the search space hypercubes so far and use them to locate the swarm particles where every swarm particle has its objective functions as coordinates.
6. Initialize each particle memory and save them in the repository as the best particle positions found so far.
7. Estimate the new velocity of every particle by using the equation:

$$\text{VEL}_i(t+1) = w\text{VEL}_i(t) + c_1 r_1(t)(\text{POP}_{i,\text{pbest}}(t) - \text{POP}_i(t)) + c_2 r_2(t)(\text{REP}_{[h]}(t) - \text{POP}_i(t)) \qquad (3.34)$$

where w represents the inertia weight, r_1 and r_2 represent random numbers in the range $[0,1]$, $POP_{i,pbest}$ represents the best position of the swarm particle i, $REP_{[h]}$ represents a value taken from the repository with index h. Roulette-wheel selection is performed to select the hypercube from which the swarm particle will be taken. Then a swarm particle is chosen randomly from the selected hypercube. POP_i represents the current fitness value of the swarm particle.

8. Estimate the new positions of each particle by using the equation:

$$POP_i(t+1) = POP_i(t) + VEL_i(t) \tag{3.35}$$

9. Maintain the swarm particles within the search space in case of going out of boundaries.

10. Evaluate the swarm particles in POP_i.

11. Update the values stored in REP and the coordinates within the hypercubes by adding the current non-dominated solutions. When the repository space is full the swarm particles in the less populated regions have more priority compared to the swarm particles in highly populated areas.

12. Update each swarm particle personal best position in the memory when the current position is better than the particle position in the memory by applying Pareto dominance, by using the equation:

$$POP_{i,\text{pbest}}(k+1) = POP_i(k) \tag{3.36}$$

13. If the termination condition is met (E.g., the maximum number of iterations), then stop the algorithm, otherwise go to step 7.

Geometric PSO

Geometric PSO (GPSO) proposed by Moraglio et al. (2007). Geometric PSO algorithm has been used in the literature similarly to the standard PSO and also for feature selection (Namous et al., 2020; Yeoh et al., 2015). GPSO differs from the standard PSO algorithm in several points: there is no velocity, the equation of position update is a convex combination and mutation is utilized. The basic equation which is utilized in GPSO is expressed by the formula:

$$x_i = CX\ ((x_i, w_1),\ (g, w_2),\ (x_i, w_3)) \tag{3.37}$$

where:

- x_i expresses the position of particle i.
- g indicates the global best, the best swarm position found so far.
- w_1, w_2, and w_3 express the scalar coefficients.
- CX expresses a randomized convex combination.

The parameters w_1, w_2, and w_3 are correlated based on the equation:

$$w_1 + w_2 + w_3 = 1 \qquad (3.38)$$

with the condition:

$$w_1, w_2, w_3 > 1 \qquad (3.39)$$

w_1 coefficient is calculated by the equation: $w_1 = 1 - w_2 - w_3$. These coefficients represent the attraction to the previous position of the particle, the particle's best position (personal learning), and the swarm best position (global learning).

Geometric PSO uses a three-parent mask-based crossover operator instead of the velocity parameter to calculate the new position of the particle (Moraglio et al., 2007). The geometric crossover guides the swarm particles in the search space in a way that they cannot be outside the convex hull of the initial swarm population. The requirements for a convex combination of w_1, w_2, and w_3 in a metric space are (Moraglio et al., 2007): Convex weights: a convex combination where $w_1, w_2, w_3 > 0$ and $w_1 + w_2 + w_3 = 1$, convexity: a convex combination returns a point within the metric convex hull, coherence between weights and distances, symmetry: when the same values are assigned to the w_1, w_2, or w_3 parameters the effect is same. The Geometric PSO algorithm is described as follows (Moraglio et al., 2007).

Geometric PSO algorithm

```
for each swarm particle i do
    initialize position xᵢ randomly in the search space
end for
while stopping criterion is not met do
    for each swarm particle i do
        set personal best xᵢ as the best particle position found so far
        set global best g as the best swarm position found so far
    end for
    for each swarm particle i do
        update particle position xᵢ using a randomized convex combination:
        xᵢ = CX ((xᵢ,wᵢ), (g,w₂), (xᵢ,w₃))
        mutate xᵢ
    end for
end while
```

PSO in neural network optimization

PSO algorithm has been applied in several researches as an artificial neural network (ANN) optimizer. The main reason for applying PSO in the ANN optimization is its advantages over other ANN optimization methods: the PSO does not require gradient information, the computation ability is improved compared to other optimization techniques such as genetic algorithm and also the PSO algorithm requires very few parameters to be defined compared to other computational optimization methods (Chen et al., 2006; Hassan et al., 2005; Kouziokas, 2020; Ou and Lin, 2006).

PSO as weight optimizer

PSO can be implemented to optimize several parameters of a neural network. An objective function should be found when a neural network is trained by PSO. The most common objective function that is used is an error metric. For example, in regression problems the mean squared error (MSE) or root mean squared error (RMSE) or mean absolute error (MAE) of a neural network can be used as an objective function. In classification problems other metrics can be utilized, such as: accuracy, precision, recall, F1, kappa. After the selection of the objective function, the swarm topology should be developed. Each particle represents a solution to the optimization of the objective function and the neural network weight vector is optimized until the swarm finds the best solution for the objective function.

PSO as topology optimizer

PSO can be implemented to optimize several parameters of a neural network such as the number of hidden neurons. In this case, each particle represents the number of hidden neurons in the hidden layer of the neural network. The objective function should be set for the PSO. As it was mentioned above, the most common objective function that is used is an error metric. Adequately, in regression problems and in classification problems several error metrics can be used as objective functions (Zhang and Shao, 2000).

According to several researches, the particle swarm optimization has a better performance in artificial neural network optimization than other optimization algorithms, such as genetic algorithm (Anand and Suganthi, 2020; Moayedi et al., 2019), imperialist competitive algorithm and artificial bee colony (Koopialipoor et al., 2019), and non-dominated sorting genetic algorithm II (NSGA-II) (Anand and Suganthi, 2020; Nourbakhsh et al., 2011). Koopialipoor et al. (2019), have studied the forecasting of a safety factor by applying several optimization techniques in artificial neural networks such as: imperialist competitive algorithm, the artificial bee colony and particle swarm optimization. The experimental outcomes have shown that PSO has a better optimization ability on the ANN compared to the other optimization algorithms. Nourbakhsh et al. (2011), have studied the application of two optimization methods in the group method of data handling neural network optimization: PSO and non-dominated sorting genetic algorithm II. The PSO optimized neural network had a better performance compared to the neural network that was optimized with the NSGA-II algorithm.

Swarm optimization examples

Some example applications of PSO in convex and non-convex optimization are illustrated in this section, regarding function approximation. The experimental set up of the PSO algorithm was applied on convex, non-convex, unimodal and multimodal functions such as: Sphere, Griewank Function, Ackley, Schwefel, Rastrigin. In all the experiments, the chosen dimension was 30. Several particle swarm architectures were investigated with different values of the acceleration coefficients c_1 and c_2. The generation (iteration) number was set to 1000.

Sphere function

Mathematical equation

The equation that represents the Sphere function is as follows:

$$f(\mathbf{x}) = f(x_1, x_2, ..., x_n) = \sum_{i=1}^{n} x_i^2 \qquad (3.40)$$

Description

The Sphere function is:

- Continuous.
- Convex.

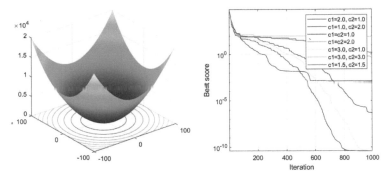

3.3a. Sphere function minimization.

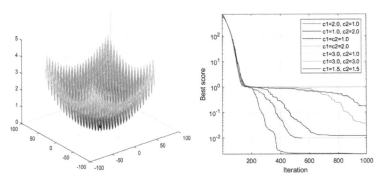

3.3b. Griewank function minimization.

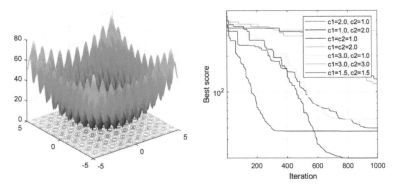

3.3c. Rastrigin function minimization.

Figure 3.3 contd. ...

...Figure 3.3 contd.

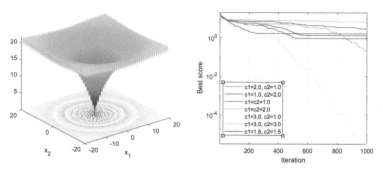

3.3d. Ackley function minimization.

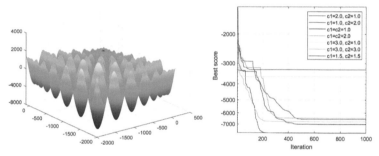

3.3e. Schwefel function minimization.

Figure 3.3. Benchmark functions minimized by PSO by testing different architectures. The particles traverse the search space in order to minimize the functions. The population size was set to 20.

- Unimodal.
- Differentiable.

Input search space

The function is evaluated with $x_i \in [-5.12, 5.12]$ with $i = 1, 2, ..., n$.

Global minima

$f(x^*) = 0$, $x^* = (0, ..., 0)$.

Function approximation with PSO

Several PSO architectures were tested. The best optimization results were achieved when the coefficient $c_1 = 2.0$ and coefficient $c_2 = 1.0$ as illustrated in Figure 3.3.

Griewank function

Mathematical equation

The equation that represents the Griewank function is as follows.

$$f(\mathbf{x}) = f(x_1,...,x_n) = 1 + \sum_{i=1}^{n} \frac{x_i^2}{4000} - \prod_{i=1}^{n} \cos(\frac{x_i}{\sqrt{i}}) \qquad (3.41)$$

Description

The Griewank function is:

- Continuous.
- Non-convex.
- Unimodal.
- Differentiable.

Input search space

The function is evaluated with xi ∈ [−600, 600] with i = 1, 2, ..., n.

Global minima

f(x*) = 0, x* = (0,...,0).

Function approximation with PSO

Several PSO architectures were tested. The best optimization results were achieved when the coefficient $c_1 = 1.0$ and coefficient $c_2 = 1.0$ as illustrated in Figure 3.3.

Rastrigin function

Mathematical equation

The equation that represents the Rastrigin function is as follows.

$$f(x) = 10d + \sum_{i=1}^{d}[x_i^2 - 10\cos(2\pi x_i)] \qquad (3.42)$$

where *d* represents the *dimension*.

Description

The Rastrigin function is:

- Continuous.
- Convex.
- Multimodal.
- Differentiable.

Input search space

The function is evaluated with xi $\in$ [−5.12, 5.12] , i = 1, 2, . . . , n.

Global minima

x* = (0, …, 0), f(x*) = 0.

Function approximation with PSO

Several PSO architectures were tested. The best optimization results were achieved when the coefficient c_1 = 1.0 and coefficient c_2 = 2.0 as illustrated in the Figure 3.3. Thus, the personal cognition is smaller than the social cognition.

Ackley function

Mathematical equation

The equation that represents the Ackley function is the following.

$$f(\mathbf{x}) = f(x_1,...,x_n) = -a.exp(-b\sqrt{\frac{1}{n}\sum_{i=1}^{n}x_i^2}) - exp(\frac{1}{n}\sum_{i=1}^{n}cos(cx_i)) + a + exp(1) \quad (3.43)$$

where a, b and c are constants which are set to the most common values for the experiments: a = 20, b = 0.2 and c = 2π.

Description

The Ackley function is:

- Continuous.
- Non-convex.
- Multimodal.
- Differentiable.

Input search space

The function is evaluated with xi $\in$ [−20, 20], and i = 1, 2, . . . , n.

Global minima

x* = (0, ..., 0), f(x*) = 0.

Function approximation with PSO

Several PSO architectures were tested. The best optimization results were achieved when the coefficient c_1 = 3.0 and coefficient c_2 = 1.0 as illustrated in the Figure 3.3. Thus, the personal cognition is bigger than the social cognition.

Schwefel function

Mathematical equation

The equation that represents the Schwefel function is the following.

$$f(\mathbf{x}) = -\sum_{i=1}^{n} x_i sin(\sqrt{|x_i|})$$

(3.44)

Description

The Schwefel function is:

- Continuous.
- Non-convex.
- Multimodal.1
- Non-differentiable.

Input search space

The function is evaluated in the search space: $x_i \in$ [−500, 500] with i = 1, 2, ..., n.

Global minima

$f(x) = -n \cdot 418.9829$, $x(i) = 420.9687$.

Function approximation with PSO

Several PSO architectures were tested. The best optimization results were achieved when the coefficient c_1 = 1.0 and coefficient c_2 = 2.0 as illustrated in the Figure 3.3. Thus, the personal cognition is smaller than the social cognition.

References

Almasi, O. N. and Khooban M. H. (2018). A parsimonious SVM model selection criterion for classification of real-world data sets via an adaptive population-based algorithm. *Neural Computing and Applications,* 30(11): 3421–3429.

Alvarez-Benitez, J. E., Everson, R. M. and Fieldsend, J. E. (2005). A MOPSO algorithm based exclusively on pareto dominance concepts. pp. 459–473. *In*: Coello Coello, C. A., Hernández Aguirre, A. and Zitzler, E. (eds.). EMO 2005. LNCS, vol. 3410. Springer, Heidelberg. https://doi.org/10.1007/978-3-540-31880-4_32.

Anand, A. and Suganthi, L. (2020). Forecasting of electricity demand by hybrid ANN-PSO models. pp. 865–882. *In*: *Deep Learning and Neural Networks: Concepts, Methodologies, Tools, and Applications*, IGI Global.

Andrews, P. S. (2006). an investigation into mutation operators for particle swarm optimization. *IEEE International Conference on Evolutionary Computation,* Vancouver, BC, Canada, 2006, pp. 1044–1051. doi: 10.1109/CEC.2006.1688424.

Assarzadeh, Z. and Naghsh-Nilchi, A. R. (2015). Chaotic particle swarm optimization with mutation for classification. *Journal of Medical Signals and Sensors,* 5(1): 12–20, Jan–Mar, 2015.

Bastos-Filho, C. J. A., Caraciolo M. P., Miranda P. B. C. and Carvalho D. F. (2008). Multi-ring particle swarm optimization. *2008 10th Brazilian Symposium on Neural Networks,* 26–30 Oct. 2008.

Bratton, D. and Kennedy, J. (2007). Defining a standard for particle swarm optimization. pp. 120–127. *In*: *Proceedings of the IEEE Swarm Intelligence Symposium*. IEEE Computer Society. https://doi.org/10.1109/SIS.2007.368035.

Chen, H. l., Yang, B., Wang, S. j., Wang, G., Liu, D. y. et al. (2014). Towards an optimal support vector machine classifier using a parallel particle swarm optimization strategy. *Applied Mathematics and Computation,* 239: 180–197. doi:https://doi.org/10.1016/j.amc.2014.04.039.

Chen, Y., Yang, B. and Dong, J. (2006). Time-series prediction using a local linear wavelet neural network, *Neurocomputing,* 69: 449–465.

Clerc, M. (2006). Confinements and biases in particle swarm optimisation. https://hal.archives-ouvertes.fr/hal-00122799/.

Coello, C. C. and Lechuga, M. S. (2002). MOPSO: A proposal for multiple objective particle swarm optimization. pp. 1051–1056. In Proceedings of the 2002 Congress on Evolutionary Computation. CEC'02 (Cat. No. 02TH8600) (Vol. 2). IEEE.

Coello, C. A. C., Pulido, G. T. and Lechuga, M. S. (2004). Handling multiple objectives with particle swarm optimization. *IEEE Transactions on Evolutionary Computation,* 8(3): 256–279.

Dasgupta, D. and Michalewicz, Z. (eds.). (2013). Evolutionary Algorithms in Engineering Applications. Springer Science & Business Media.

de Carvalho, D. F. and Bastos-Filho, C. J. A. (2009). Clan particle swarm optimization. *International Journal of Intelligent Computing and Cybernetics*.

Deb, K. (2011). Multi-objective optimisation using evolutionary algorithms: an introduction. pp. 3–34. In Multi-objective Evolutionary Optimisation for Product Design and Manufacturing. Springer, London.

Eberhart, R. C. and Shi, J. Y. (2001). Kennedy, *Swarm Intelligence*: Elsevier.

Eberhart, R. and Kennedy, J. (1995). A new optimizer using particle swarm theory. *MHS'95. Proceedings of the Sixth International Symposium on Micro Machine and Human Science* [4–6 Oct. 1995].

Esquivel, S. C. and Coello, C. C. (2003). On the use of particle swarm optimization with multimodal functions. pp. 1130–1136.

Hassan, R., Cohanim, B., de Weck, O. and Venter, G. (2005). A comparison of particle swarm optimization and the genetic algorithm. pp. 1–13. *In: 46th AIAA/ASME/ASCE/ AHS/ASC Structures, Structural Dynamics and Materials Conference, American Institute of Aeronautics and Astronautics.*

Helwig, S., Branke, J. and Mostaghim, S. (2013). Experimental analysis of bound handling techniques in particle swarm optimization. *IEEE Trans. Evol. Comput.* 17(2): 259–271. https://doi.org/10.1109/TEVC.2012.2189404.

Higashi, N. and Iba H. (2003). Particle swarm optimization with gaussian mutation. pp. 72–79. *In: Proceedings of the IEEE Swarm Intelligence Symphosium 2003.* IEEE Press.

Huang, T. and Mohan, A. S. (2005). A hybrid boundary condition for robust particle swarm optimization. *IEEE Antennas Wirel. Propag. Lett.* 4: 112–117. https://doi. org/10.1109/LAWP.2005.846166.

Kennedy, J. (2002). Article swarm performance. pp. 1671–1676. In Proceedings of the 2002 Congress on Evolutionary Computation. CEC'02 (Cat. No. 02TH8600) (Vol. 2). IEEE.

Koopialipoor, M. Armaghani, D. J. Hedayat, A. Marto, A., Gordan, B et al. (2019). Applying various hybrid intelligent systems to evaluate and predict slope stability under static and dynamic conditions. *Soft Computing*, 23: 5913–5929.

Kouziokas, G. N. (2020). A new W-SVM kernel combining PSO-neural network transformed vector and Bayesian optimized SVM in GDP forecasting. *Engineering Applications of Artificial Intelligence*, 92: 103650. doi:https://doi.org/10.1016/j. engappai.2020.103650.

Lin, Z. and Zhang, Q. (2017). An effective hybrid particle swarm optimization with Gaussian mutation. *Journal of Algorithms & Computational Technology*, 11(3): 271–280.

Michalewicz, Z. (1998). *Genetic Algorithms + Data Structures = Evolution Programs*: Springer Science & Business Media.

Miranda, V., Keko, H. and Junior, A. J. (2008). Stochastic star communication topology in evolutionary particle swarms (EPSO).

Moayedi, H. Moatamediyan, A. Nguyen, H., Bui, X.-N., Bui, D. T. et al. (2019). Prediction of ultimate bearing capacity through various novel evolutionary and neural network models. *Engineering with Computers*, 1–17.

Moraglio, A., Chio, C. D. and Poli, R. (2007, April). Geometric particle swarm optimisation. pp. 125–136. *In: European Conference on Genetic Programming.* Springer, Berlin, Heidelberg.

Mostaghim, S., Mostaghim, S., Halter, W. and Wille, A. (2006). Linear multi-objective particle swarm optimization. pp. 209–238. *In: Ajith, A., Crina, G. and Vitorino, R. (eds.). Stigmergic Optimization*, vol. 31. Springer, Heidelberg. https://doi. org/10.1007/978-3-540-34690-6_9.

Namous, F., Faris, H., Heidari, A. A., Khalafat, M., Alkhawaldeh, R. S. and Ghatasheh, N. (2020). Evolutionary and swarm-based feature selection for imbalanced data classification. pp. 231–250. *In*: *Evolutionary Machine Learning Techniques*. Springer, Singapore.

Nourbakhsh, A. Safikhani, H. and Derakhshan, S. (2011). The comparison of multi-objective particle swarm optimization and NSGA II algorithm: applications in centrifugal pumps. *EnOp*, 43: 1095–1113.

Ou, C. and Lin, W. (2006). Comparison between PSO and GA for parameters optimization of PID controller. pp. 2471–2475. *In*: *2006 International Conference on Mechatronics and Automation, IEEE*.

Padhye, N., Deb, K. and Mittal, P. (2013). Boundary handling approaches in particle swarm optimization. pp. 287–298. *In*: Bansal, J. C., Singh, P. K., Deep, K., Pant, M. and Nagar, A. K. (eds.). *Proceedings of Seventh International Conference on Bio-Inspired Computing: Theories and Applications*, vol. 1. Springer, India (2013). https://doi.org/10.1007/978-81-322-1038-2_25.

Robinson, J. and Rahmat-Samii, Y. (2004). Particle swarm optimization in electromagnetics. *IEEE Trans. Antennas Propag.* 52(2): 397–407. https://doi.org/10.1109/TAP.2004.823969.

Sabat, S. L., Udgata, S. K. and Murthy, K. P. N. (2010). Small signal parameter extraction of MESFET using quantum particle swarm optimization. *Microelectronics Reliability*, 50(2): 199–206, 2010/02/01/.

Shi, Y. and Eberhart, R. (1998). *A Modified Particle Swarm Optimizer*. Paper presented at the 1998 IEEE International Conference On Evolutionary Computation Proceedings. IEEE World Congress on Computational Intelligence (Cat. No. 98TH8360).

Stacey, A., Jancic, M. and Grundy, I. (2003). Particle swarm optimization with mutation. *IEEE Congress on Evolutionary Computation*, Dec. 2003, pp. 1425–1430.

Sun, J., Feng, B. and Xu, W. (2004). Particle swarm optimization with particles having quantum behavior. pp 325–331. *In*: *Proceedings of the IEEE Congress on Evolutionary Computation 2004 (IEEE CEC'04), Portland (OR)*.

Sun, J, Lai, C.-H. and Wu, X.-J. (2011). *Particle Swarm Optimisation: Classical and Quantum Perspectives*. CRC Press, Boca Raton.

Tian, Z. (2021). Modes decomposition forecasting approach for ultra-short-term wind speed. *Applied Soft Computing*, 107303. doi:https://doi.org/10.1016/j.asoc.2021.107303.

Van den Bergh, F. and Engelbrecht, A. P. (2006). A study of particle swarm optimization particle trajectories. *Information Science*, 176(8): 937–971.

Yeoh, T. W., Zapotecas-Martínez, S., Akimoto, Y., Aguirre, H. E. and Tanaka, K. (2015, September). Feature selection in gait classification using geometric PSO assisted by SVM. pp. 566–578. *In*: *International Conference on Computer Analysis of Images and Patterns*. Springer, Cham.

Yin, P.-Y., Glover, F., Laguna, M. and Zhu, J.-X. (2011). A complementary cyber swarm algorithm. *International Journal of Swarm Intelligence Research (IJSIR)*, 2(2): 22–41.

Zhang, C. and Shao, H. (2000). An ANN's evolved by a new evolutionary system and its application. *Proceedings of the 39th IEEE Conference on Decision and Control (Cat. No. 00CH37187)*, 4: 3562–3563.

Zhang, W. J., Xie, X. F. and Bi, D. C. (2004). Handling boundary constraints for numerical optimization by particle swarm flying in periodic search space. pp. 2307–2311. *In*: *Proceedings of Congress on Evolutionary Computation*.

Chapter 1

Ant Colony Optimization and Artificial Bee Colony

◇◇◇

This chapter is devoted to the most important colony-based optimization algorithms: Ant colony optimization (ACO) and Artificial Bee Colony (ABC). The standard Ant colony optimization (ACO) is presented and some important ant-based algorithms are explained such as: Ant System, Ant Colony System, Rank-based Ant System, Max-min Ant System and Population-based ACO. Then the standard Artificial Bee Algorithm is presented by explaining also the bee colony foraging behavior. Furthermore, advanced aspects are discussed, such as: ABC selection methods including: disruptive selection, tournament selection, exponential and linear rank-based selection and also, boundary handling approaches which can restrict the artificial bee colony movement in the predefined search space.

Ant colony optimization (ACO)

Introduction

Ant colony optimization algorithm was inspired by the natural foraging behavior of ants. The ants communicate with each other in order to find the shortest path between their nest and the food resources by using pheromone trails. This is the main idea that was used to build an ant colony-based optimization algorithm in order to solve several kinds of optimization problems (Chandra Mohan and Baskaran, 2012; Colorni et al., 1991; Dorigo, 1992; Dorigo et al., 1996; Dorigo et al., 2006; Dorigo and Blum, 2005; Dorigo and Gambardella, 1997; Blum, 2005; López-Ibáñez et al., 2018). Several modifications have been developed in order to improve the performance of

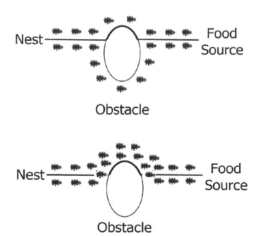

Figure 4.1. Illustration of the ant foraging behavior. The ant colony will find the shortest path from the nest to the food communicating with each other by using pheromone trails.

the originally proposed ACO algorithm such as, chaotic random spare ACO (Zhao et al., 2021), knowledge-based ACO (Xing et al., 2010), parallel ACO (Pedemonte et al., 2010), Rank-Based Version of the Ant system (Bullnheimer et al., 1997), hyper-cube framework ACO (Blum et al., 2001), population-based ACO (Scheuermann et al. 2004).

ACO was initially proposed for discrete optimization problems, such as the travelling salesman problem and routing problems. Later, ant colony optimization was also proposed for continuous optimization problems (Socha and Dorigo, 2008). For example, ACO was used to train a feed-forward neural network (Socha and Blum, 2007). Furthermore, ACO algorithm was applied on several kinds of problems and domains such as, routing in ad hoc wireless networks (Sharvani et al., 2012), automatic pressurized water reactor loading pattern design (Lin and Lin, 2012), construction site layout planning (Lam et al., 2007), solving symmetric and asymmetric TSPs (Gambardella and Doringo, 1996), strongly constrained optimization (Maniezzo and Milandri, 2002).

ACO algorithm use pheromone updates, evaporation, and daemon (optional). ACO can optimize a problem by using an updated pheromone trail and directing the artificial ants in the search space according to the transition probability and the total pheromone in the area.

The continuous ant colony optimization (Socha and Dorigo, 2008) applies both local and global search. The local artificial ants move toward the area with the best solution according to the transition probability of area a, expressed by the equation:

$$P_a(t) = \frac{t_a(t)}{\sum_{j=1}^{n} t_j(t)} \qquad (4.1)$$

where,

- $t_a(t)$ represents the total pheromone in a area
- n represents the total number of ants.

The pheromone is updated by using the following mathematical equation:

$$t_i(t+1) = (1-r)t_i(t) \qquad (4.2)$$

where r represents the pheromone evaporation rate.

In the following sections, the most important ant-based algorithms are examined.

Ant system (AS)

The Ant System algorithm was proposed by Colorni et al. (1991). This approach uses locally interacting simple agents called ants which deposit pheromone when they travel from a node i (Or town in the Travelling Salesman Problem) to a node j ().

The intensity on the (i,j) trail path according to the amount of pheromone is defined by the equation:

$$\tau_{ij}(t) = (1-\rho)\tau_{ij}(t-1) + \Delta\tau_{ij}(t) \qquad (4.3)$$

- where $\tau_{ij}(t)$ represents the amount of pheromone on the (i,j) path at time t
- ρ denotes the evaporation coefficient and $0 < \rho < 1$
- $\Delta\tau_{ij}(t)$ denotes the increase in the pheromone amount on the (i,j) path between time instants (t−1) and t.

The $\Delta\tau_{ij}(t)$ represents the total pheromone increase by all the ants which have travelled in this time interval and is calculated by the equation:

$$\Delta\tau_{ij}(t) = \sum_{k} \Delta\tau_{ijk}(t) \qquad (4.4)$$

where $\Delta\tau_{ijk}(t)$ is the amount of pheromone deposited by the k-th ant on the (i,j) path in the time step.

The transition probability regarding the choice from a node i (e.g., town in the TSP) to another node j, by an ant at node i at time t is defined by the equation:

$$p_{ij}(t) = \frac{\left[\tau_{ij}(t)\right]^{\alpha}\left[\eta_{ij}\right]^{\beta}}{\sum\limits_{z\in N_i}\left[\tau_{il}(t)\right]^{\alpha}\left[\eta_{iz}\right]^{\beta}}$$

(4.5)

where,

- α and β are parameters used to let the user to set the relative importance of trail versus visibility

- η_{ij} represents the *visibility* parameter, the inverse of the distance between node i and node j, calculated as: $\eta_{ij} = \dfrac{1}{d_{ij}}$

The visibility parameter η_{ij} was introduced in order to enhance the efficiency of the optimization algorithm by enabling the nearest nodes to be selected with a higher probability than the distant nodes.

Colorni et al. (1991), proposed three different methods for calculating the value of $\Delta_{\tau ijk}(t)$, named as ANT-density, ANT-quantity and ANT-cycle. In the Ant Density and the Ant Quantity models the pheromone update is applied when an ant moves from one location to another. In the Ant Cycle model, the pheromone update is applied when all ants have finished their tour.

In the ANT-quantity model (Colorni et al., 1991) a constant pheromone quantity Q_1 is left when the ant travels from i to j and in the ANT-density model when an ant travels from i to j leaves pheromone quantity units Q_2 for each unit of length. The value of the $\Delta\tau_{ijk}(t)$ is given by the equation:

$$\Delta\tau_{ijk}(t) = \begin{cases} \dfrac{Q_1}{d_{ij}}, & \text{if } (i,j)\in T_k, \\ 0, & \text{otherwise,} \end{cases}$$

(4.6)

where d_{ij} is the distance between i and j, T_k represents the path finished by the k-th ant and Q_1 is a constant. In the Ant quantity model, the path length distance d_{ij} between i and j is taken into account.

In the *Ant density model*, the Q_2 constant value, shows that the final pheromone left on the path is proportional to the number of ants selecting

70

it. This model doesn't take into account the path length distance d_{ij}. The value of $\Delta\tau_{ijk}(t)$ is given by the equation:

$$\Delta\tau_{ijk}(t) = \begin{cases} Q_2, & \text{if } (i,j) \in T_k, \\ 0, & \text{otherwise.} \end{cases}$$
(4.7)

In the *ant cycle model (Or online delayed)*, the pheromone update is implemented when all the ants have finished their tour.

The updating of the trail value of the next time step $\tau_{ij}(t + 1)$ can be calculated by the equation:

$$\tau_{ij}(t+1) = \rho\tau_{ij}(t) + \Delta\tau_{ijk}(t)$$
(4.8)

The value of $\Delta\tau_{ijk}(t)$ after a complete tour is calculated by:

$$\Delta\tau_{ijk}(t) = \begin{cases} \dfrac{Q_3}{L_k}, & \text{if } (i,j) \in T_k \\ 0, & \text{otherwise} \end{cases}$$
(4.9)

where,

- T_k represents the path finished by the k-th ant
- L_k represents the path length
- Q_3 represents a constant.

The ANT-cycle model has showed the best performance compared to the other models.

The ant system algorithm is summarized in the following pseudo code.

Initialize the parameters of the ant system

For t = 1 to N do

   For k = 1 to m do

      Repeat until ant k has completed a tour

         Select the node j to be visited next with probability p_{ij} calculated by eq. (4.5)

         Calculate the length L_k of the tour produced by the ant k

      Update the trail values τ_{ij} on all paths according to equation (4.8)

End

---

where N represents the number of total iterations.

Ant colony system (ACS)

The basic differences of Ant colony system compared to the Ant system are (Dorigo and Gambardella, 1997): the state transition rule is altered in order to accomplish a balance between exploring new paths compared to the already accumulated ant knowledge, the local pheromone updating rule is calculated when artificial ants create a tour and the global pheromone updating rule is applied only to paths of the best tour.

State transition rule

The state transition rule is developed by using two different strategies, exploitation and exploration. The exploitation strategy always selects the path with the greatest amount of pheromone visibility and intensity. The exploration process is a stochastic rule according to Eq. 4.5. When an ant decides which node (Or city) j to visit next, it chooses the node (Or city) to move to defined by the following rule:

$$j = \begin{cases} \arg\max_{z \in N_i} \tau_{iz}(t) \cdot (\eta_{iz})^{\alpha}, & \text{if } q \leq q_0 \text{ (exploitation)} \\ S, & \text{otherwise (exploration)} \end{cases} \quad (4.10)$$

where,

- q_0 ($0 \leq q_0 \leq 1$) represents a pre-defined variable which expresses the distribution ratio of the two strategies
- q represents a uniformly and randomly generated value distributed in the interval $[0,1]$
- S expresses the transition probability rule calculated by the Eq. 4.5.

Global pheromone updating rule

In the Ant colony system, in the optimization process, only the ant with the globally best value is permitted to deposit pheromone on the path. This makes the ant search more focused: ants search in the near neighborhood of the best tour found in the current iteration. The Global updating is implemented when all the ants have completed their tours in the search space. The pheromone amount is updated by using the following updating rule:

$$\tau_{ij}(t) = (1-a) \cdot \tau_{ij}(t-1) + a \cdot \Delta\tau_{ij}(t) \quad (4.11)$$

where,

$$\Delta \tau_{ij} = \begin{cases} \left(L_{Global}\right)^{-1}, & \text{if } (i,j) \in T_G \\ 0, & \text{otherwise} \end{cases} \qquad (4.12)$$

where,

- $0 < \alpha < 1$ represents the global pheromone decay parameter
- L_{Global} represents the length of the globally best ant tour from the initial start point.

Local pheromone updating rule

After finishing a tour, the pheromone level is updated on each ant path according to the equation:

$$\tau_{ij}(r,s) = (1-\rho) \cdot \tau_{ij}(r,s) + \rho \cdot \Delta \tau_{ij}(r,s) \qquad (4.13)$$

where,

- $0 < r < 1$ represents the local decay rate parameter.
- ρ denotes the evaporation coefficient rate and $0 < \rho < 1$.

The value of $\Delta \tau_{ij}(t)$ is either specified based on the results of Q-learning or assigned the value of the initial pheromone level.

Rank-based ant system (RB-AS)

Rank-based Ant System (RB-AS) was proposed by Bullnheimer et al. (1999). It basically consists of an ant-based solution development and pheromone update in every iteration.

According to the proposed method, during the solution search, an artificial ant assembles the solution as a sequence of solution components by using probabilistic theory. At every step, the selection of a solution from a set of probable solutions is produced by using the transition probability which is calculated by the equation:

$$p(i,s) = \frac{[\tau(i,s)]^\alpha [\eta(i,s)]^\beta}{\sum_k [\tau(i,k)]^\alpha [\eta(i,k)]^\beta} \forall s \in N(s) \qquad (4.14)$$

where,

- $p(i,s)$ represents the probability of choosing a solution for position i from the solution set
- $N(s)$, contains the solutions to be placed at each solution development step

- $\tau(i, s)$ represents the pheromone concentration of solutions at position i
- $\eta(i, s)$ represents an optional weighting function.
- α and β represent the weights of the pheromone concentration (τ) and heuristic information (η).

The pheromone update rule is composed of two parts, the pheromone evaporation and the pheromone increase. The pheromone evaporation decreases all the pheromone concentrations. The best solution is the one with the highest pheromone concentration. The pheromone increase permits the solutions from the current or earlier iterations to be used to increase the solution of the component pheromone concentration.

The pheromone update can be calculated by the following equation:

$$\tau(i,s) \leftarrow (1-\rho)\tau(i,s) + \rho \sum_{l \in S_{update}} \Delta\tau_l(i,s) \forall s \in N(s) \tag{4.15}$$

$$\Delta\tau_l(i,s) = \begin{cases} \dfrac{w_l}{Q_l} & \text{if } (i,s) \in l \\ 0 & \text{if } (i,s) \notin l \end{cases} \tag{4.16}$$

where,
- $\tau(i, s)$ represents the pheromone concentration,
- $0 < \rho \leqslant 1$ represents the evaporation rate,
- Q_l represents the quality function value,
- w_l represents the weighting factor of the quality function value,
- S_{update} represents a solution set used to update the solution components' pheromone.

Max-min ant system (MMAS)

Max-min Ant System was proposed by Stützle and Hoos (2000). It is based on the idea of limiting the range of the pheromone in the main ant algorithm. The artificial ants deposit pheromones on the optimal route or on the optimal route in the current iteration but this can lead to an excessive growth of pheromones. To avoid this excessive growth, Max-min Ant System limits the range of values of pheromones within the interval of the lower and the upper bound $[\tau_{lb}, \tau_{ub}]$. Consequently, the Max-Min Ant System algorithm, updates the pheromones according to the equation:

$$\tau_{i,j} = \left[(1-\rho)\tau_{i,j} + \Delta\tau_{i,j}^{best} \right]_{\tau_{lb}}^{\tau_{ub}} \qquad (4.17)$$

where,

- $\Delta\tau_{i,j}^{best}$ represents the total quantity of pheromone deposited on the i, j route by the optimal ant
- τ_{ub} represents the upper bound of pheromones
- τ_{lb} represents the lower bound of pheromones,

The $[x]_{\tau_{lb}}^{\tau_{ub}}$ is estimated by the following equation:

$$[x]_{\tau_{lb}}^{\tau_{ub}} = \begin{cases} ub & x > ub \\ lb & x < lb \\ x & \text{otherwise} \end{cases} \qquad (4.18)$$

The total quantity of pheromone deposited by the best ant is estimated by the following equation:

$$\Delta\tau_{ij}^{best} = \begin{cases} \dfrac{1}{L_{best}} & \text{if}\,(i, j) \in \text{Best-solution} \\ 0 & \text{otherwise} \end{cases} \qquad (4.19)$$

where L_{best} represents the distance of the best route.

The initial level of the pheromone is set to τ_{max} to make the artificial ants search for multiple optimal routes at the initial stage. Also, when the algorithm does not find a better path for a predefined number of iterations, the pheromone levels are reinitialized so as to reduce the prematurity potential risk.

Population-based ACO

In regular ACO, the solutions found by the artificial ants are discarded when the pheromone update is implemented. In the PB-ACO algorithm the solutions are kept and eventually the solutions are removed according to a rule (age, quality, prob, elitism) and the pheromone values are updated in the table.

The population update strategies proposed by Guntch (2004) and Guntch and Middendorf (2002) are summarized as: the age-based strategy, the quality-based strategy, the prob strategy and the elitist-based strategy.

Age-based strategy. In this strategy the oldest solution is removed after k iterations and a new candidate solution replaces it. This means that every solution has an influence for exactly k iterations.

Quality-based strategy. In this strategy, the best k solutions found for all past iterations are stored (not only the best solutions of the last k iterations). If the solution of a current iteration is better than the previous best then it is replaced.

Prob strategy. The prob strategy improves the *quality* strategy by introducing a probabilistic selection of which element from P unified with the candidate solution will be discarded and will not be part of P in the next algorithm iteration. Otherwise, the artificial ants would be trapped to search in a small section of the search space.

Elitism strategy. In this strategy, elitism is implemented on the population based ACO algorithm, so the update strategy used by the algorithm will be applied only to the population of the elitist solutions (best solutions found so far). The elitist solution is updated only when a better candidate solution is found.

Artificial bee colony (ABC)

Bee colony foraging behavior

Several researchers have studied the honeybee colony foraging behavior. They have developed a model by expressing this behavior by the reaction–diffusion equations (Tereshko, 2000; Tereshko and Lee, 2002; Tereshko and Loengarov, 2005). This model which expresses the collective intelligence of honeybee swarms is composed of three essential factors: food sources, employed foragers, and unemployed foragers. Tereshko and Loengarov (2005) explain these three essential factors of their proposed model:

Food Sources: The food source is very valuable to an insect. The criteria of the forager bee for selecting a food source are basically: how close is the food source to the nest, concentration of energy and how difficult it is for the bee to extract this energy. Tereshko and Loengarov (2005) group all these parameters with the term "profitability" of a food source, a single quantity parameter.

Employed foragers: The employed forager bees are related to a specific food source which they are currently exploiting. The employed foragers carry information about this specific source, regarding the distance and the direction from the nest, and also about the profitability of the food source. The employed forager bees share this information with a specific probability which depends on the profitability of the food source. If the profitability of the food source is bigger, then the probability of the honeybee to do a waggle dance and share the information will be higher.

Unemployed foragers: The unemployed foragers are searching for food sources in order to exploit them. There are two types of unemployed forager bees, the scout bees, who are responsible for searching the closer environment around the nest for new food sources and onlooker bees just wait in the nest and find a new food source according to the information shared by the employed forager bees.

ABC algorithm

The artificial bee colony (ABC) algorithm was proposed by Karaboga in 2005. Since then, several variations of the original ABC algorithm have been proposed such as, Chaotic ABC (Xu et al., 2010), ABC for clustering (Zhang et al., 2010), tournament selection based ABC algorithm with elitist strategy (Zhang et al., 2014), global best-guided ABC (Zhu and Kwong, 2010), improved ABC (Kang et al., 2010), ABC algorithm based on Boltzmann selection policy (Ding and Feng, 2009), ABC with new versions of onlooker bee (Awadallah et al., 2019). The artificial bee colony can be used to find solutions in constrained and unconstrained optimization problems (Karaboga and Basturk, 2007; Sulaiman et al., 2012). In the artificial bee colony (ABC) algorithm, the optimization process was inspired by the intelligence of the natural foraging behavior of bee colonies. In the proposed ABC algorithm, three kinds of artificial bees are employed to search for food sources, employed bee, onlooker bee, and scout bee. Correspondingly, there are three stages according to the bee kinds,

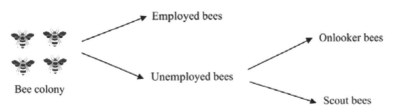

Figure 4.2. Artificial bee colony taxonomy.

employed bee stage, onlooker bee stage, and scout bee stage. A food source represents a candidate solution to an optimization problem.

Initialization stage: The ABC algorithm begins the search process by setting an initial bee population randomly distributed SN solutions (food source places). SN represents the size of the population. Each food source represents a solution and is initialized by using the following equation:

$$x_{i,j} = x_j^{min} + rand(0,1) \cdot \left(x_j^{max} - x_j^{min} \right) \tag{4.20}$$

where,

- $x_{i,j}$ represents the solution for the i_{th} employed artificial bee in the j_{th} dimension
- $X_1 = (x_{i,1}, x_{i,2},..., x_{i,D})$ represents the i_{th} food source,
- D represents the dimension size of the optimization problem.
- $[x_j^{min}, x_j^{max}]$ represents the upper and lower bounds for the j_{th} dimension.
- $rand(0, 1)$ represents a random number uniformly distributed in the range [0, 1].

Employed bee stage. In this stage, the swarm of employed bees are searching for food resources in order to make honey. The employed artificial bees create a new food source $v_{i,j}$ (new solution) for each $x_{i,j}$ parent food source according to the following equation:

$$v_{i,j} = x_{i,j} + \phi_{i,j} \cdot \left(x_{i,j} - x_{k,j} \right) \tag{4.21}$$

where,

- $v_{i,j}$ is a new solution for
- $x_{i,j}$ represents the solution for the i_{th} employed artificial bee in the j_{th} dimension
- j represents a randomly selected dimension from the set [1,...D].
- $\phi_{i,j}$ represents a random number uniformly distributed in the range [−1, 1].

The set $V_i = (v_{i,1}, v_{i,2},..., v_{i,D})$ represents the ith new food source. $X_k = (x_{k,1}, x_{k,2},..., x_{k,D})$ represents a parent food source randomly chosen.

The fitness function value for a food source is estimated by the following equation:

$$fitness(X_i) = \begin{cases} \dfrac{1}{1+f(X_i)} & \text{if} \quad f(X_i) \geq 0 \\ 1+|f(X_i)| & \text{otherwise} \end{cases} \qquad (1.22)$$

where,

- *fitness(X$_i$)* is the fitness function value of the food source.
- *f(X$_i$)* is a specific objective function.
- *|f(X$_i$)|* denotes the absolute value of the specific objective function.

Onlooker bee stage. In this stage, a swarm of SN onlooker bees is sent to select a food source based on the its quality. If the quality of food sources is high then they attract more onlooker bees, otherwise if the quality is low then the onlooker bees are not attracted. The quality of food sources is determined according to the fitness function of the solutions. The onlooker bees receive information by the employed bees about the food sources, and then the onlooker bees choose the high-quality food sources to search and produce new food sources (new solutions) based on a selection method (e.g., roulette selection method). The selection probability is calculated by the equation:

$$p_i = \frac{fitness(X_i)}{\sum\limits_{j=1}^{SN} fitness(X_j)} \qquad (4.23)$$

where,

- p_i is the selection probability of the ith solution for the food source
- *fitness(X$_i$)* is the fitness value of the ith solution

Scout bee stage. The role of scout bees is to reinitialize the food sources. When a food source has not been improved for an amount of time which can be set as a threshold value parameter, then the food source should be abandoned and the scout bees are searching for new food sources. The abandoned food source (solution) is replaced by a randomly created new food source (new solution) by using the equation (4.20).

Selection methods

The selection method of the individuals according to their fitness values in the search space is very important in the ABC algorithm. The most suitable selection method can improve the efficiency of the algorithm (Bao and Zeng, 2009; Kumar et al., 2016).

Disruptive selection

This method uses a normalized-by-mean fitness function in order to improve the population diversity (Kuo and Hwang, 1996). The selection probability is calculated by the equation:

$$p_i = \frac{fit_i}{\sum_{j=1}^{N} fit_j} \qquad (4.24)$$

where,

- fit_i, represents the fitness value of the ith solution
- P_i represents the selection probability of the ith solution.

The fitness function is calculated by the equation:

$$fit_i = \left| f_i - \overline{f} \right| \qquad (4.25)$$

where,

- f_i is the objective function
- $\overline{f}$ is the average of the objective values for the individuals in the population.

This method was implemented by several researches to improve the optimization results (Alzaqebah and Abdullah, 2011; Karaboga, 2009). For example, Alzaqebah and Abdullah (2011), have used disruptive selection to calculate the selection probability of the onlooker bees.

Tournament selection

This selection method randomly selects k individuals from the population, where k is taken as the tournament size and the fitness values of the individuals are compared. A score is assigned to the one with the best fitness value. The individuals are chosen according to the probability calculated by the following equation:

$$p_i = \frac{S_i}{\sum_{i=1}^{k} S_i} \tag{4.26}$$

where,

- S represents the score assigned to the individual with the best fitness value
- k represents the number of individuals from the population

Linear rank selection

In this method, the individuals are ranked according to their fitness values. They are ranked from the worst fitness value to the best fitness value. The selection probability is calculated according to the equations (Aiguo and Jiren, 1999; Bao and Zeng, 2009):

$$P_k = \frac{1}{n} + a(t)\frac{n+1-2k}{n(n+1)} \qquad k = (1,2,\cdots,n) \tag{4.27}$$

$$a(t) = 0.2 + \frac{3t}{4N} \qquad t = (1,2,\cdots,N) \tag{4.28}$$

where,

- $a(t)$ represents a self-adaptive parameter
- N represents the maximum iterations.

Exponential rank selection

In this method, the individuals are ranked according to the exponential weighing of ranked individuals and the selection probability is calculated according to the equation:

$$p_i = \frac{c-1}{c^N - 1}c^{N-i}, i\epsilon\{1,\ldots,N\} \tag{4.29}$$

where indicative parameter c < 1.

Boundary handling approaches

Several boundary handling approaches have been proposed. These techniques are very important since they can restrict the artificial bee colony movement in the predefined search space. In the previous chapter several boundary handlings have been presented in detail for PSO algorithm such as: nearest or boundary or absorb, hyperbolic method, infinity or invisible wall, random, random-half, periodic, exponential, mutation, reflect methods, random damping. These methods can also be applied in the ABC algorithm.

Furthermore, there are other approaches that have been proposed by researchers for the ABC algorithm. Brajevic (2015), proposed the application of a mechanism used by Kukkonen and Lampinen (2006) in generalized differential evolution. Brajevic (2015), used the Kukkonen and Lampinen (2006) mechanism as a boundary constraint-handling method. During the algorithm process of the mutation and crossover, the trial vector $u_{j,i,G}$ is compared with the old vector $x_{j,i,G}$. This correction is based on the following mathematical equation which is used in the case of boundary violation values which were reflected back from the violated boundary by using following rule:

$$u_{j,i,G} = \begin{cases} 2 \cdot x_j^{low} - u_{ji}, & \text{if } u_{j,i,G} < x_j^{low} \\ 2 \cdot x_j^{up} - u_{ji}, & \text{if } u_{j,i,G} > x_j^{up} \\ u_{j,i,G}, & \text{otherwise} \end{cases} \tag{4.30}$$

where,

- $u_{j,i,G}$ represents the trial vector for the variable j of the candidate solution i,
- x_j^{low} represents the lower bound of the decision variable x_i
- x_j^{up} represents the upper bound of the decision variable x_j
- G represents the generation

References

Aiguo, S. and Jiren, L. (1999). A ranking based adaptive, evolutionary operator genetic algorithm [J]. *ACTA, Electronica Sinica*, 27(1): 85–88.

Alzaqebah, M. and Abdullah, S. (2011). *Hybrid Artificial Bee Colony Search Algorithm Based on Disruptive Selection for Examination Timetabling Problems.* Paper presented at the Combinatorial Optimization and Applications, Berlin, Heidelberg.

Awadallah, M. A., Al-Betar, M. A., Bolaji, A. L. a., Alsukhni, E. M. et al. (2019). Natural selection methods for artificial bee colony with new versions of onlooker bee. *Soft Computing*, 23(15): 6455–6494. doi:10.1007/s00500-018-3299-2.

Bao, L. and Zeng, J. C. (2009, August). Comparison and analysis of the selection mechanism in the artificial bee colony algorithm. pp. 411–416. *In: 2009 Ninth International Conference on Hybrid Intelligent Systems* (Vol. 1). IEEE.

Blum, C., Roli, A. and Dorigo, M. (2001). HC–ACO: The hyper-cube framework for ant colony optimization. pp. 399–403. In: *Proceedings of MIC'2001—Metaheuristics International Conference* (Vol. 2). Porto, Portugal, July 16–20, 2001.

Blum, C. (2005). Ant colony optimization: Introduction and recent trends. *Physics of Life Reviews*, 2(4): 353–373. doi:https://doi.org/10.1016/j.plrev.2005.10.001.

Brajevic, I. (2015). Crossover-based artificial bee colony algorithm for constrained optimization problems. *Neural Computing and Applications*, 26(7): 1587–1601. doi:10.1007/s00521-015-1826-y.

Bullnheimer, B., Hartl, R. F. and Strauss, C. (1997). *A New Rank-Based Version of the Ant System—A Computational Study*. Technical report. Austria: Institute of Management Science, University of Vienna, Austria.

Bullnheimer, B., Hartl, R. F. and Strauss, C. (1999). An improved ant System algorithm for the vehicle Routing Problem. *Annals of Operations Research*, 89: 319–328.

Chandra Mohan, B. and Baskaran, R. (2012). A survey: Ant colony optimization based recent research and implementation on several engineering domains. *Expert Systems with Application*, 39: 4618–4627.

Colorni, A., Dorigo, M. and Maniezzo, V. (1991, December). Distributed optimization by ant colonies. pp. 134–142. In: *Proceedings of the First European Conference on Artificial Life* (Vol. 142).

Ding, H. J. and Feng, Q. X. (2009). Artificial bee colony algorithm based on Boltzmann selection policy. *Computer Engineering and Applications*, 45(31): 53–55.

Dorigo, M. (1992). *Optimization, Learning, and Natural Algorithms*. Ph.D. Thesis. Politecnico di Milano, Milan, Italy.

Dorigo, M., Maniezzo, V. and Colorni, A. (1996). Ant system: optimization by a colony of cooperating agents. *IEEE Transactions on Systems, Man, and Cybernetics, Part B (Cybernetics)*, 26(1): 29–41.

Dorigo, M. and Gambardella, L. M. (1997). Ant colony system: a cooperative learning approach to the traveling salesman problem. *IEEE Transactions on Evolutionary Computation*, 1(1): 53–66.

Dorigo, M. and Blum, C. (2005). Ant colony optimization theory: A survey. *Theoretical Computer Science*, 344(2-3): 243–278.

Dorigo, M., Birattari, M. and Stuzle, T. (2006). Ant Colony Optimisation. Artificial Ants as a computational Intelligence Technique. Technical Report, IRDIA, 1–12, 1781-3794, 2006.

Dorigo, M., Birattari, M. and Stutzle, T. (2006). Ant colony optimization. *IEEE Computational Intelligence Magazine*, 1(4): 28–39.

Gambardella, L. M. and Dorigo, M. (1996, May). Solving symmetric and asymmetric TSPs by ant colonies. pp. 622–627. In: *Proceedings of IEEE International Conference on Evolutionary Computation*. IEEE.

Gao, W., Liu, S. and Huang, L. (2012). A global best artificial bee colony algorithm for global optimization. *Journal of Computational and Applied Mathematics*, 236(11): 2741–2753.

Guntsch, M. and Middendorf, M. (2002, April). A population-based approach for ACO. pp. 72–81. In: *Workshops on Applications of Evolutionary Computation*. Springer, Berlin, Heidelberg.

Guntsch, M. and Middendorf, M. (2002, September). Applying population based ACO to dynamic optimization problems. pp. 111–122. *In*: *International Workshop on Ant Algorithms*. Springer, Berlin, Heidelberg.

Guntsch, M. (2004). *Ant Algorithms in Stochastic and Multi-criteria Environments* (Doctoral dissertation, Karlsruhe Institute of Technology).

Kang, F., Li, J., Li, H., Ma, Z., Xu, Q. et al. (2010, May). An improved artificial bee colony algorithm. pp. 1–4. *In*: *2010 2nd International Workshop on Intelligent Systems and Applications*. IEEE.

Karaboga, D. (2005). *An Idea Based on Honey Bee Swarm for Numerical Optimization* (Vol. 200, pp. 1–10). Technical report-tr06, Erciyes University, Engineering Faculty, Computer Engineering Department.

Karaboga, D. and Basturk, B. (2007). A powerful and efficient algorithm for numerical function optimization: artificial bee colony (ABC) algorithm. *Journal of Global Optimization*, 39(3): 459–471.

Karaboga, N. (2009). A new design method based on artificial bee colony algorithm for digital IIR filters. *Journal of the Franklin Institute*, 346(4): 328–348.

Kukkonen, S. and Lampinen, J. (2006, July). Constrained real-parameter optimization with generalized differential evolution. pp. 207–214. *In*: *2006 IEEE International Conference on Evolutionary Computation*. IEEE.

Kumar, A., Kumar, D. and Jarial, S. K. (2016). A comparative analysis of selection schemes in the artificial bee colony algorithm. *Computación y Sistemas*, 20(1): 55–66.

Kuo, T. and Hwang, S. Y. (1996). A genetic algorithm with disruptive selection. *IEEE Transactions on Systems, Man, and Cybernetics, Part B* (Cybernetics), 26(2): 299–307.

Lam, K. C., Ning, X. and Ng, T. (2007). The application of the ant colony optimization algorithm to the construction site layout planning problem. *Construction Management and Economics*, 25(4): 359–374. doi:10.1080/01446190600972870.

Lin, C. and Lin, B.-F. (2012). Automatic pressurized water reactor loading pattern design using ant colony algorithms. *Annals of Nuclear Energy*, 43: 91–98. doi:https://doi.org/10.1016/j.anucene.2011.12.002.

López-Ibáñez, M., Stützle, T. and Dorigo, M. (2018). Ant colony optimization: a component-wise overview. pp. 371–407. *In*: Martí, R., Pardalos, P. M. and Resende, M. G. C. (Eds.), *Handbook of Heuristics*. Cham: Springer International Publishing.

Maniezzo, V. and Milandri, M. (2002). An ant-based framework for very strongly constrained problems. *In*: Dorigo, M., Di Caro, G. and Sampels, M. (eds.). *Proceedings of ANTS 2002—From Ant Colonies to Artificial ANTS: Third International.*

Pedemonte, M., Nesmachnow, S. and Cancela, H. (2011). A survey on parallel ant colony optimization. *Applied Soft Computing*, 11(8): 5181–5197.

Scheuermann, B., So, K., Guntsch, M., Middendorf, M., Diessel, O. et al. (2004). FPGA implementation of population-based ant colony optimization. *Applied Soft Computing*.

Sharvani, G. S., Ananth, A. G. and Rangaswamy, T. M. (2012). Analysis of different pheromone decay techniques for ACO based routing in ad hoc wireless networks. *International Journal of Computer Applications*, 56(2).

Socha, K. and Blum, C. (2007). An ant colony optimization algorithm for continuous optimization: application to feed-forward neural network training. *Neural Computing and Applications*, 16(3): 235–247.

Socha, K. and Dorigo, M. (2008). Ant colony optimization for continuous domains. *European Journal of Operational Research*, 185(3): 1155–1173.

Stützle, T. and Hoos, H. H. (2000). MAX–MIN ant system. *Future Generation Computer Systems*, 16(8): 889–914.

Sulaiman, M. H., Mustafa, M. W., Shareef, H. and Khalid, S. N. A. (2012). An application of artificial bee colony algorithm with least squares support vector machine for real and reactive power tracing in deregulated power system. *International Journal of Electrical Power & Energy Systems*, 37(1): 67–77.

Tereshko, V. (2000, September). Reaction-diffusion model of a honeybee colony's foraging behaviour. pp. 807–816. *In: International Conference on Parallel Problem Solving from Nature*. Springer, Berlin, Heidelberg.

Tereshko, V. and Lee, T. (2002). How information-mapping patterns determine foraging behaviour of a honey bee colony. *Open Systems & Information Dynamics*, 9(2): 181–193.

Tereshko, V. and Loengarov, A. (2005). Collective decision making in honey-bee foraging dynamics. *Computing and Information Systems*, 9(3): 1.

Xing, L. N., Chen, Y. W., Wang, P., Zhao, Q. S., Xiong, J. et al. (2010). A knowledge-based ant colony optimization for flexible job shop scheduling problems. *Applied Soft Computing*, 10(3): 888–896.

Xu, C., Duan, H. and Liu, F. (2010). Chaotic artificial bee colony approach to Uninhabited Combat Air Vehicle (UCAV) path planning. *Aerospace Science and Technology*, 14(8): 535–541.

Zhang, C., Ouyang, D. and Ning, J. (2010). An artificial bee colony approach for clustering. *Expert Systems with Applications*, 37(7): 4761–4767.

Zhang, M. D., Zhan, Z. H., Li, J. J. and Zhang, J. (2014, November). Tournament selection based artificial bee colony algorithm with elitist strategy. pp. 387–396. *In: International Conference on Technologies and Applications of Artificial Intelligence*. Springer, Cham.

Zhao, D., Liu, L., Yu, F., Heidari, A. A., Wang, M., Liang, G. et al. (2021). Chaotic random spare ant colony optimization for multi-threshold image segmentation of 2D Kapur entropy. *Knowledge-Based Systems*, 216: 106510. doi:https://doi.org/10.1016/j.knosys.2020.106510.

Zhu, G. and Kwong, S. (2010). Gbest-guided artificial bee colony algorithm for numerical function optimization. *Applied Mathematics and Computation*, 217(7): 3166–3173.

CHAPTER 5

Cuckoo Search and Bat Swarm Algorithm

◇◇◇

This chapter is devoted to the swarm optimization algorithms: Cuckoo Search (CS) and Bat Algorithm (BA). The standard Cuckoo Search based on the Lévy Flights and some important variations of the Cuckoo Search algorithm are presented such as: Chaotic Cuckoo Search, Discrete Cuckoo Search and Discrete Binary Cuckoo Search. Also, the Bat Algorithm and some important variations of the algorithm are discussed such as: Chaotic Bat Algorithm, Binary Bat Algorithm, Discrete Bat Algorithm and also two recent variants: Bat Algorithm with double mutation (2020) and Self-adaptive bat algorithm (2019).

Cuckoo search

Cuckoo search is a meta-heuristic swarm algorithm proposed by Yang and Deb (2009) for several kinds of continuous optimization problems. The algorithm is focused on the natural behavior of the cuckoo species.

Cuckoo breeding behavior and Lévy flights

Cuckoo birds are well known for their aggressive reproduction strategy and for their beautiful sounds. Some of the cuckoo species like *ani* and *Guira* cuckoos are communal breeders. Cuckoo birds are communally nesting birds, and usually female birds lay their eggs into a shared—common nest and then provide parental care cooperatively in a mixed clutch (Grieves and Quinn, 2018).

They lay their eggs in nests and they are capable of discriminating among parasitic and non-parasitic eggs (Hauber et al., 2018; Macedo, 1992; Payne and Sorensen, 2005). Cuckoo birds are known for their breeding habits which are parasitic, taking into consideration that female birds lay eggs in host nests and abandon their parenthood responsibilities. Most cuckoo species remove the host eggs from the nest in a few days after hatching (Hughes, 1996). There are several types of brood parasitisms such as: intraspecific, cooperative breeding, and nest take over.

Several studies have shown that the flight behavior of many birds and insects is characterized by Lévy flights (Brown et al., 2007; Pavlyukevich, 2007; Reynolds and Frye, 2007). This natural behavior has already been idealized and applied on computational optimization and the results are very promising (Boudjemaa et al., 2020; Sharma et al., 2013; Zhou et al., 2018).

Cuckoo search algorithm

Yang and Deb (2009) developed three idealized rules regarding the description of the behavior of the Cuckoo species:

- Each cuckoo lays one egg each time and then dumps its laid egg in a randomly selected nest.
- The best nests with the optimal egg quality will carry over to the following generations.
- The number of available cuckoo host nests is constant, and the laid egg is discovered by the cuckoo host bird with a probability $p_a \in [0, 1]$. A fraction p_a of the cuckoo nests is substituted by the new ones.

Yang and Deb (2009, 2010) used simple representations, every egg in the cuckoo nest represents a solution, and a new laid cuckoo egg represents a new solution but the proposed algorithm can be extended to the more complex optimization problems by defining that each bird nest has many eggs representing a whole set of solutions.

The basic steps of the Cuckoo Search (CS) algorithm as proposed by Yang and Deb (2009) is expressed by the next pseudo code:

Cuckoo Search algorithm via Levy Flights

```
begin

    Objective function f (x), x = (x₁, ..., xₐ)ᵀ

    Generate the initial bird population of n host nests xᵢ (i = 1, 2, ..., n)

    while (t < MaximumGeneration) or (other stopping criterion)

        Get a cuckoo randomly using Lévy flights and evaluate its fitness function Fitᵢ

        Select j nest between n host nests randomly

        if (Fitᵢ > Fitⱼ),

            replace j nest by the new solution of i nest;

        end

        A fraction pₐ of the worse bird nests is abandoned and the new ones are built;

        Keep the optimal solutions;

        Rank the solutions found and find the current best

    end while

    Postprocess the results and visualize them

End
```

The Cuckoo search algorithm utilizes a combination of a local random walk (exploitation) and a global random walk (exploration) in a balanced way, and it is controlled by the variable *Pa*. The local random walk (exploitation) is expressed by the equation:

$$\mathrm{x}_i^{(t+1)} = \mathrm{x}_i^{(t)} + \alpha s \otimes H(Pa - \in) \otimes (\mathrm{x}_j^{(t)} - \mathrm{x}_k^{(t)}) \tag{5.1}$$

where,

- $x_j^{(t)}$ and $x_k^{(t)}$ represent two separate solutions randomly selected.
- $H(u)$ represents a Heaviside function.

- $\in$ represents a random number uniformly distributed in [0,1].
- s represents the step size.
- $\otimes$ represents the entry-wise product.
- *Pa* represents a variable for balancing the exploitation and exploration.

During the algorithm process, when generating new solutions x(*t*+1) in time step t+1 for a cuckoo *i*, then the Lévy flight is implemented by using the following equation:

$$x_i^{(t+1)} = x_i^{(t)} + \alpha \otimes \text{Levy}(\lambda) \tag{5.2}$$

where,

- $\alpha > 0$ represents the step size which is related to the problem. In most problems, the value of α is close to 1.
- The product $\otimes$ represents the entry wise multiplication.

This equation is basically the stochastic equation used in random walk. This entry wise product is similar to those used in PSO algorithm. The Lévy flight is considered more efficient for exploring the search space in each step.

The Lévy flight basically provides a random walk and the random step length is expressed by the Lévy distribution

$$\text{Levy} \sim u = t^{-\lambda}, \quad (1 < \lambda \leq 3) \tag{5.3}$$

This distribution has an infinite variance and infinite mean. The steps basically are from a random walk. Several new solutions are generated by Lévy walk close to the best solution so far which speed ups the search at a local level. Although several new solutions are randomly generated in locations far from the current best solution, this facilitates the process so that it will not be trapped in a local best solution.

The Cuckoo Search is a population-based optimization algorithm which uses elitism and selection methods similar to the harmony search, the randomization is considered as more efficient since the step length can be also very large and the number of parameters that have to be tuned is smaller than the genetic algorithm and particle swarm optimization.

Cuckoo search variants

Chaotic cuckoo search

Wang et al. (2016), proposed a chaotic cuckoo search optimization algorithm by incorporating the chaos theory into the cuckoo search algorithm. Also, twelve chaotic maps were tested to tune the step size of the cuckoos in the original search algorithm. Twenty-seven benchmark functions were used to investigate the efficiency of the proposed chaotic cuckoo search algorithm.

The chaos has the property that a value cannot be repeated and the search speeds are higher compared to stochastic searches which are based on probabilities. The main parameters of the standard cuckoo search algorithm which have an increased influence on the convergence performance are: the step size called α and the discovery rate p_a. The authors have used the value 0.25 for parameter *pa* for the standard cuckoo search algorithm according to the literature (Yang, 2010b).

In the standard cuckoo search algorithm, the step size remains constant. On the contrary in the chaotic cuckoo search algorithm the step size α *is a* chaotic variable which can accelerate the algorithm convergence. Furthermore, all chaotic maps are normalized in the interval [0, 2]. The pseudo-code of chaotic CS algorithm proposed by Wang et al. (2016) is the following:

Chaotic Cuckoo Search algorithm

---

begin

  Initialization: Set the counter t=1; initialize population P randomly;

    set p_a initial value of the chaotic map c_o randomly and the elitism parameter K

  while (t < MaximumGeneration) **do**

    Sort the population according to their fitness.

    Store the K best cuckoos.

    Update the step size using chaotic maps ($\alpha = c_{t+1}$).

    Get an i cuckoo randomly and substitute its solution by using Lévy flights.

```
Evaluate its fitness Fit_i.

Choose a nest j among n randomly

if (Fit_i > Fit_j),

    replace j by the new solution;

end if

A fraction p_a of the worse bird nests is abandoned and the new ones are built;

Replace the K worst cuckoo solution with the K best cuckoo solution.

Sort the bird population and find the current best.

t = t+1

end while

End
```

The following chaotic maps were used by Wang et al., 2016 to test the performance of the chaotic CS:

a) Chebyshev map

$$x_{k+1} = \cos(k \cos^{-1}(x_k)) \qquad (5.4)$$

b) Circle map which generates a chaotic sequence in the interval (0, 1), with a = 0.5 and b = 0.2,

$$x_{k+1} = x_k + b - (a/2\pi)\sin(2\pi k) \bmod(1) \qquad (5.5)$$

c) Gaussian map

$$x_{k+1} = \begin{cases} 0 & x_k = 0 \\ 1/x_k \bmod(1) & \text{otherwise} \end{cases}, 1/x_k \bmod(1) = \frac{1}{x_k} - \left[\frac{1}{x_k}\right] \qquad (5.6)$$

d) Intermittency map

$$x_{k+1} = \begin{cases} \varepsilon + x_k + cx_k^n & 0 < x_k \leq P \\ \dfrac{x_k - P}{1 - P} & P < x_k < 1 \end{cases} \qquad (5.7)$$

e) Iterative map

$$x_{k+1} = \sin\left(\frac{a\pi}{x_k}\right), \quad a \in (0,1) \tag{5.8}$$

f) Liebovitch map

$$x_{k+1} = \begin{cases} \alpha x_k & 0 < x_k \leq P \\ \dfrac{P - x_k}{P_2 - P_1} & P_1 < x_k \leq P_2, \\ 1 - \beta(1 - x_k) & P_2 < x_k \leq 1 \end{cases} \tag{5.9}$$

g) Logistic map

$$x_{k+1} = ax_k(1 - x_k) \tag{5.10}$$

h) Piecewise map

$$x_{k+1} = \begin{cases} \dfrac{x_k}{P} & 0 \leq x_k < P \\ \dfrac{x_k - P}{0.5 - P} & P \leq x_k < \dfrac{1}{2} \\ \dfrac{1 - P - x_k}{0.5 - P} & \dfrac{1}{2} \leq x_k < 1 - P \\ \dfrac{1 - x_k}{P} & 1 - P \leq x_k < 1 \end{cases} \tag{5.11}$$

i) Sine map

$$x_{k+1} = \frac{a}{4}\sin(\pi x_k), \quad 0 < a \leq 4 \tag{5.12}$$

j) Singer map

$$x_{k+1} = \mu(7.86x_k - 23.31x_k^2 + 28.75x_k^3 - 13.302875x_k^4) \tag{5.13}$$

k) Sinusoidal map

$$x_{k+1} = ax_k^2 \sin(\pi x_k) \tag{5.14}$$

I) Tent map

This chaotic map is similar to the logistic map and is expressed by the following equation:

$$x_{k+1} = \begin{cases} \dfrac{x_k}{0.7} & x_k < 0.7 \\ \dfrac{10}{3} & x_k \geq 0.7 \end{cases} \tag{5.15}$$

According to the results in several benchmark functions, Wang et al. (2016), have concluded that the chaotic CS algorithm was improved with the incorporation of the chaotic sinusoidal map instead of α. Also, they have concluded that the tuned CS algorithm improves the global search ability.

Discrete binary cuckoo search

Gherboudj et al. (2012) proposed a discrete binary cuckoo search optimization algorithm that can be used in binary optimization problems. The solutions are represented as a set of real numbers in a continuous search space (Continuous optimization) or by an integer set in a discrete search space (discrete optimization). Some discrete optimization problems are: the job shop scheduling (Pongchairerks, 2009), and the flow shop scheduling problems (Liao et al., 2007).

The standard CS algorithm for continuous optimization problems is based on *Lévy* flights. Consequently, the algorithm solutions constitute a set of real numbers in a continuous search space. In order to use CS in discrete binary problems the solutions should be converted to binary values. The binary CS is composed of the following basic modules:

- The Levy flights which are used to obtain a new cuckoo solution.
- The binary solution representation used to compute the flipping chances for every cuckoo by using the sigmoid function.
- The objective function and the selection operator.

In the conversion process from a continuous to a binary area, if x_i is a continuous solution in the space [0, 1] and x_i' is a binary solution representation, the sigmoid function is used to convert the solution values according to the following equation:

$$S(x_i) = \frac{1}{1+e^{-x_i}}$$

(5.16)

where $S(x_i)$ is the flipping chance of the binary solution $\dot{x}_i$.

In order to calculate the binary solution, $S(x_i)$ is compared to the value of the random number generated in the interval [0, 1] for every dimension i of the solution x according to the following equation:

$$x_i' = \begin{cases} 1 & \gamma << S(x_i) \\ 0 & otherwise \end{cases}$$

(5.17)

Where γ represents the random number in the space [0,1].

Hybrid self-adaptive cuckoo search

Mlakar et al. (2016), proposed a hybrid self-adaptive cuckoo search algorithm by modifying some mechanisms of the original cuckoo search algorithm. The authors proposed balancing the exploration techniques, self-adapting the cuckoo search parameters, and adapting the cuckoo population.

Balance to the exploration strategies

An efficient balancing between the exploration and the exploitation strategies is proposed by setting the algorithm parameters appropriately. The exploration can be classified into three categories according to the distance: the randomly guided long-distance, short-distance and moderate-distance explorations. The first strategy regards the global random walk, the second strategy exploits the neighborhood by using the local random walk and the third exploration strategy can be found in the differential evolution algorithm (Storn and Price, 1997) and has two parameters: the distance from the current best solution and the distance from two solutions randomly selected.

The proposed differential evolution mutation strategy can be expressed by the following equation:

$$\mathbf{u}_i^{(t)} = \mathbf{x}_i^{(t)} + F_i\left(\mathbf{x}_{best}^{(t)} - \mathbf{x}_i^{(t)}\right) + \left(\mathbf{x}_{r1}^{(t)} - \mathbf{x}_{r2}^{(t)}\right), \tag{5.18}$$

where,

- F_i represents a scaling factor regarding the measure of change
- $\mathbf{x}_{best}^{(t)}$ represents the current best solution
- $\mathbf{x}_{r1}^{(t)}$ and $\mathbf{x}_{r2}^{(t)}$ represent the solutions selected randomly from the population.

The authors proposed a crossover operator which can be expressed by the following equation:

$$w_{i,j}^{(t)} = \begin{cases} u_{i,j}^{(t)} & rand_j\,(0,1) \le CR \vee j = j_{rand}, \\ x_{i,j}^{(t)} & otherwise, \end{cases} \tag{5.19}$$

where,

- $CR \in [0.0, 1.0]$ represents the crossover rate which controls the rate of the parameters copied to the new solution.
- $j = j_{rand}$ represents a condition which guarantees that the new vector differs from the original solution $\mathbf{x}_i^{(t)}$ at least in one element.

The substitution of the solution which is randomly selected can be expressed by the following equation:

$$\mathbf{x}_k^{(t+1)} = \begin{cases} \mathbf{w}_i^{(t)} & if\ f\left(\mathbf{w}_i^{(t)}\right) \le f\left(\mathbf{x}_k^{(t)}\right) \wedge k \ne i \\ \mathbf{x}_i^{(t)} & otherwise, \end{cases} \tag{5.20}$$

where, $k = rand(0, P)$ represents a randomly selected number uniformly distributed in the range $[0, P)$ and P represents the population size parameter.

The following equation was proposed for balancing the search strategies controlled by the balancing probability factor p_b in the hybrid self-adaptive cuckoo search algorithm:

$$if \begin{cases} U(0,1) \le p_b, & then\ moderate\ distance\ strategy \\ otherwise, & then\ short\ distance\ strategy \end{cases} \tag{5.21}$$

where $U(0, 1)$ represents a random value uniformly produced between 0 and 1.

Self-adaptation of the control parameters

Mlakar et al. (2016), proposed a self-adaptation strategy for the parameters of the cuckoo search. The solutions are represented by the following equation:

$$\mathbf{x}_i^{(t)} = (x_{i,1}^{(t)},\ldots,x_{i,D}^{(t)},\alpha_i^{(t)},\lambda_i^{(t)},F_i^{(t)},CR_i^{(t)})^T, \quad for \; i=1,\ldots,P. \quad (5.22)$$

where,

- α represents the scaling factor for the short-distance exploration.
- λ represents the stability factor for the short-distance exploration.
- F represents scaling factor for the moderate-distance exploration.
- CR represents crossover rate for the moderate-distance exploration.
- P represents the population size.

The parameters of the short-distance exploration search strategy are self-adapted based on the following equations:

$$\alpha_i^{(t+1)} = \begin{cases} r_1, & if \; r_2 < \tau_0, \\ \alpha_i^{(t)}, & otherwise, \end{cases} \quad (5.23)$$

$$\lambda_i^{(t+1)} = \begin{cases} r_3, & if \; r_4 < \tau_1, \\ \lambda_i^{(t)}, & otherwise, \end{cases} \quad (5.24)$$

$$F_i^{(t+1)} = F_i^{(t)}, \quad CR_i^{(t+1)} = CR_i^{(t)}, \quad (5.25)$$

The moderate-distance self-adaptation is performed by the equation:

$$F_i^{(t+1)} = \begin{cases} r_5, & if \; r_6 < \tau_2, \\ F_i^{(t)}, & otherwise, \end{cases} \quad (5.26)$$

$$CR_i^{(t+1)} = \begin{cases} r_7, & if \; r_8 < \tau_3, \\ CR_i^{(t)}, & otherwise, \end{cases} \quad (5.27)$$

$$\alpha_i^{(t+1)} = \alpha_i^{(t)}, \quad s_i^{(t+1)} = s_i^{(t)}, \quad (5.28)$$

where $\tau_0,\ldots,\tau_3$ represent the learning rates when the self-adaptive parameters change, and $r_1,\ldots,r_s$ represent random numbers distributed uniformly within the range $[0,1]$ and s represents the solution in a time step.

Population reduction

Small Cuckoo populations have faster convergence but can be trapped in a local optimum, and larger populations have slower convergence but improved exploration. The authors proposed a linear population reduction. Mlakar et al. (2016), proposed the population size to be reduced based on the following equation:

$$P^{(t+1)} = round\left[\left(\frac{P_{max} - P_{min}}{MAX\_Eval}\right) * Eval + P_{min}\right], \qquad (5.29)$$

where,

- P_{max} represents the initial Cuckoo population size (maximum).
- P_{min} represents the specified minimal Cuckoo population size.
- *Eval* and *MAX_Eval* represent the current and maximum number of function evaluations, correspondingly.

Bat algorithm

Bat algorithm inspiration

There are several types of bats in nature with similarities. Bats use a sonar when navigating and hunting. They have the ability to decrease the loudness and increase the emittance rate of ultrasonic sound when hunting for prey. Yang (2010a) proposed bat-inspired algorithm idealizing some characteristics of microbats by using the following idealized rules:

1. All the bats use echo-location to sense the distance, and they 'know' the difference between food and background barriers.

2. Bats fly randomly with the velocity v_i at position x_i with a constant frequency f_{min}, with a varying wavelength λ and loudness A_0 to search for the prey. They can automatically adjust the frequency depending on the closeness of the target.

3. Although the loudness A_0 can vary in many ways, it is assumed that the loudness range is from a large A_0 value to a minimum constant value A_{min}.

The pseudocode of the algorithm according to Yang (2010a) is the following:

Bat Algorithm

Objective function $f(\mathbf{x})$, $\mathbf{x} = (x_1, ..., x_d)^T$

Initialize bat population $\mathbf{x}_i$ ($i = 1, 2, ..., n$) and $\mathbf{v}_i$

Define the pulse frequency f_i at bat population $\mathbf{x}_i$

Initialize the pulse rates r_i and loudness A_i

while *(t < MaxIterationNumber)*

    Generate the new solutions by adjusting frequency,

    and update velocities and solutions according to the equations 5.29 to 5.31

        if *(rand > r_i)*

          Select a solution between the best solutions

          Generate a local solution around the selected best solution

        end if

    Generate a new bat solution by flying randomly

    if *(rand < A_i & $f(\mathbf{x}_i) < f(\mathbf{x}_*)$)*

        Accept the new solutions

        Increase r_i and reduce A_i

    end if

    Rank the bats and find the current best $\mathbf{x}$

end while

    Postprocess results and visualization

Bat movement

In the bat algorithm the vectors with their frequencies, positions x_i, and velocities v_i in a d-dimensional search space are calculated according to the following equations (Yang, 2010a):

$$F_i = F_{min} + (F_{max} - F_{min})\beta \tag{5.30}$$

where,

- F_i represents the frequency of the i-th bat which is updated in each iteration.
- β represents a random number of a uniform distribution in $[0,1]$.

$$v_i^{(t+1)} = v_i^{(t)} + (x_i^{(t)} - x^*)F_i \tag{5.31}$$

$$x_i^{(t+1)} = x_i^{(t)} + v_i^{(t+1)} \tag{5.32}$$

where,

- $\mathbf{x}^{t+1}$ represents the new position (solution)
- $\mathbf{v}^{t+1}$ represents the new velocity
- $\mathbf{x}^*$ represents the global best solution so far

Yang (2010b) proposed a local search, when a solution is chosen among the current best positions/solutions. Consequently, a new bat solution is generated locally by using random walk according to the following equation:

$$X_{new} = X_{old} + \varepsilon A^t \tag{5.33}$$

where,

- ε represents a random number in the interval $[-1,1]$,
- A is the loudness parameter of the emitted sound that bats utilize to perform exploration instead of exploitation.

The loudness A_i and the rate r_i of the pulse emission are updated as the iterations proceed according to the following equations:

$$A_i^{(t+1)} = \alpha A_i^{(t)} \tag{5.34}$$

$$r_i^{(t+1)} = r_i^{(0)} \left[1 - \exp(-\gamma t)\right] \tag{5.35}$$

where,

- α represents a constant and represents a cooling factor similar to simulated annealing.
- γ represents a constant.

Bat algorithm variants

Several modifications of the Bat algorithm were proposed in the literature such as, binary bat algorithm (Nakamura et al., 2012), chaotic bat algorithm (Gandomi and Yang, 2014), self-adaptive bat algorithm (Lyu et al., 2019) bat algorithm with double mutation operators (Liu et al., 2020), bat algorithm for constrained optimization tasks (Gandomi et al., 2013), enhanced Bat algorithm with mutation operator (Ghanem and Jantan, 2019) and discrete bat algorithm (Luo et al., 2014).

Binary bat algorithm

Nakamura et al. (2012), have proposed a binary bat algorithm for feature selection. In the standard bat algorithm the bat swarm moves in a continuous search space. In a feature selection problem, the search space is constructed as a *n*-dimensional (n expresses the number of features) Boolean search space since it should be decided whether a feature should be selected or not.

The bat's positions (solutions) should be represented as binary vectors. Nakamura et al. (2012), have used a sigmoid function to express the bat positions in binary values:

$$S(v_i^j) = \frac{1}{1 + e^{-v_i^j}} \tag{5.36}$$

The bat position is calculated according to the following equation:

$$x_i^j = \begin{cases} 1 & if \ S(v_i^j) > \sigma \\ 0 & otherwise \end{cases} \tag{5.37}$$

where $\sigma \sim U(0,1)$. Therefore, only binary values can be provided for each bat coordinates.

According to Nakamura et al. (2012), the proposed binary bat algorithm outperformed other similar techniques used on several datasets.

Mirjalili et al. (2014), have proposed a different approach of a binary bat algorithm. Since the sigmoid function forces the bats to take the values of 0 and 1, their velocity increases but the positions remain the same.

Mirjalili et al. (2014), proposed a v-shaped transfer function for position updating to oblige the bats which have high velocity to switch their positions:

$$V\left(v_i^k(t)\right) = \left|\frac{2}{\pi} \arctan\left(\frac{\pi}{2} v_i^k(t)\right)\right| \tag{5.38}$$

$$x_i^k(t+1) = \begin{cases} \left(x_i^k(t)\right)^{-1} & \text{If} \quad \text{rand} < V\left(v_i^k(t+1)\right) \\ x_i^k(t) & \text{rand} \geq V\left(v_i^k(t+1)\right) \end{cases} \tag{5.39}$$

where $x_i^k(t)$ and v_i^k indicate the position and velocity of the i_{th} particle at iteration t in the k_{th} dimension, and $(x_i^k(t))^{-1}$ is the complement of $x_i^k(t)$.

The authors have tested their approach in twenty-two benchmark functions and the results were compared with Binary PSO and the genetic algorithm. The results have shown an increased efficiency of the proposed algorithm.

Chaotic bat algorithm

Gandomi and Yang (2014), proposed a Chaotic Bat Algorithm by using several chaotic maps in order to tune the Bat Algorithm parameters and improve the experimental results and the algorithm performance. The Chaotic maps that were used in the experiments are the following: Chebyshev map, Circle map, Gauss/Mouse map, Intermittency map, Iterative map, Liebovitch map, Logistic map, Piecewise map, Sawtooth map, Sine map, Singer map, Sinusoidal map, Tent map. Most of the equations of these chaotic maps are presented in the chaotic CS section.

The proposed chaotic Bat algorithms are the following:

1st Chaotic bat algorithm

Gandomi and Yang (2014), proposed the modification of the parameter β in the frequency update equation by using chaotic maps expressed by the following equation:

$$f_i = f_{min} + (f_{max} - f_{min})CM_i \tag{5.40}$$

where CM represents a chaotic map. In the conventional bat algorithm, β represents a random number from 0 to 1 and in the 1st chaotic version of the bat algorithm it is a chaotic number between 0 and 1 according to the selected chaotic map. The experimental results have shown an improved performance of the chaotic bat algorithm compared to the conventional bat algorithm.

2nd Chaotic bat algorithm

Also, Gandomi and Yang (2014), proposed the modification of the parameter λ_i in the velocity update equation by using chaotic maps expressed by the following equation

$$v_i^t = v_i^{t-1} + (x_i^t - x^*)CM_i f_i \tag{5.41}$$

where CM represents a chaotic map.

3rd Chaotic bat algorithm

Also, Gandomi and Yang (2014), proposed the modification of the loudness (A) parameter in the bat algorithm by replacing it with chaotic maps in order to improve the bat algorithm performance.

Self-adaptive bat algorithm

Lyu et al. (2019), proposed a self-adaptive bat algorithm by incorporating a step-control and a mutation mechanism based on the observations and the algorithm global search should be adapted to according to the loudness parameter A and the local search process should be adapted according to the pulse emission rate parameter R.

Step-control mechanism

Lyu et al. (2019), proposed a step-control mechanism to control the step sizes used in every iteration in the global and local search. The proposed algorithm utilizes two frequencies f_1 and f_2, for the individual bats and for the bat swarm, accordingly.

First, during the global search, the new solutions are expressed by the following equations:

$$v_i^t = \omega v_i^{t-1} + f_1 r_1 \left(h_{i*} - x_i^{t-1} \right) + f_2 r_2 \left(x_* - x_i^{t-1} \right) \tag{5.42}$$

$$f_1 = \alpha\left(1 - e^{-|F_{avg} - F_{best}|}\right) + \gamma(1 - k) + f_{min} \tag{5.43}$$

$$C_w = f_1 + f_2 \tag{5.44}$$

$$x_i^t = x_i^{t-1} + \mu v_i^t \tag{5.45}$$

where,

- ω represents a decreasing weight coefficient.
- h_{i*} represents the optimal position for bat i.
- x_* represents the current global optimum.
- f_1 and f_2 represent the frequencies as described above.
- r_1 and r_2 represent random numbers uniformly distributed in (0.5, 1.5).
- F_{avg} represents the average fitness of the current optimal individual positions.
- F_{best} represents the best fitness of the current global optimum solution.
- $k = t/tmax$ represents an evaluation index with $k \in (0, 1]$.
- $tmax$ represents the maximum number of iterations.
- $fmin$ represents the minimum value of f_1.
- μ represents the step weight coefficient in the range (0, 1].
- α and γ are weights.

This search is conducted according to the following equations:

If the evaluation index is: $k < 0.4$, then

$$x_i^t = x_* + A^t s\delta \times g(k) \tag{5.46}$$

If the evaluation index is: $k \geq 0.4$, then

$$x_i^t = x_* + A^t s\delta \times 0.1^{g(k)} \tag{5.47}$$

where,

- A^t represent the bat swarm current average loudness.
- $\delta \in [-1, 1]$ represent a random number uniformly distributed.
- s represents among the upper bound and the lower bound of the obtainable solution and the number of bat agents in the swarm group.
- $g(k)$ is a function related to the search steps.

Mutation mechanism

Lyu et al. (2019), proposed a step-control mechanism which improved the algorithm global search capability. Also, the authors proposed a mutation mechanism that integrates the loudness A to improve the algorithm's capability to avoid local optima.

The loudness A in the standard bat algorithm reduces gradually to 0 in the later steps, and the pulse emission rate R increases gradually to *Rmax*. The proposed equations for the updates of the A and R parameters are expressed by the equations:

$$A_i^{t+1} = \frac{f_1}{f_{max}} \tag{5.48}$$

$$R_i^{t+1} = \frac{f_2}{f_{max}} \tag{5.49}$$

where *fmax* represents the upper limit of the frequency.

The frequencies f_1 and f_2 are adaptively updated according to the bat swarm fitness and accordingly the A and R parameters are adapted according to the above equations.

Bat algorithm with double mutation

The standard bat algorithm shows a poor performance in problems with a complex nature due to its early convergence. Liu et al. (2020), proposed a bat algorithm with double mutation operators where a modified factor related to time and two mutation operators is incorporated. The proposed algorithm appears to improve the bat algorithm performance in nonlinear optimization problems according to the experimental results.

Liu et al. (2020), proposed three modifications of the standard bat algorithm by introducing: a time factor parameter, a Cauchy mutation operator and a Gaussian mutation operator.

Time factor modification

In the standard bat algorithm, the new bat positions are updated by utilizing the sum of the previous iteration and the current velocity. During the algorithm implementation the distance between the bats and the best solution globally decreases and also the bat flight time decreases.

The authors introduced a constant as a time factor in the original equation which calculates the next bat position. The time factor λ_t is close to 1 at the beginning of the algorithm in order to increase speed of the convergence, and when the algorithm arrives close to the selected number of iterations which was set as a limit the time factor λ_t is close to "0" to enhance the local search. The proposed time factor is calculated by the following equation:

$$\lambda_t = 1 - e^{\left(-\left(\frac{it_{max} - it}{it_{max} + 0.6it} \right)^{10} \right)}$$

(5.50)

where,

- it_{max} represents the maximum number of iterations.
- it represents the number of current iterations.

The new position equation with the time factor λ_t is expressed by the formula:

$$X_i^t = X_i^{t-1} + \lambda_t V_i^t$$

(5.51)

Cauchy mutation operator modification

Liu et al. (2020), proposed a third modification to the original bat algorithm by integrating a Cauchy mutation operator which can help the bat algorithm to escape from a local optimum according to the literature (Yang and Huang, 2012).

The function of the one-dimensional Cauchy density is expressed by the following equation (Paiva et al., 2017):

$$c(x) = \frac{1}{\pi} \frac{k}{k^2 + x^2}$$

(5.52)

where,

- $x \in [-\infty, \infty]$
- parameter $k > 0$ is a scale factor.

The Cauchy distribution function is expressed by the following equation (Paiva et al., 2017):

$$C_t(x) = \frac{1}{2} + \frac{1}{\pi} arctan\left(\frac{x}{k} \right)$$

(5.53)

The modifying position update equation with Cauchy mutation operator is integrated into the original bat algorithm and expressed with the following formula:

$$X_i^t = X_i^t + X_i^t Cauchy(0,1) \tag{5.54}$$

where *Cauchy* (0,1) is a random number which follows the Cauchy distribution function. The *k* parameter was set to 1.

Gaussian mutation operator

Liu et al. (2020), proposed the Gaussian mutation operator to be used in the exploitation of the algorithm in the last iterations since the Gaussian mutation operator size is smaller than the Cauchy mutation operator size.

The one-dimensional Gaussian density function is expressed by the equation:

$$g(x) = \frac{1}{\sigma\sqrt{2\pi}} e^{\left[\frac{-(x-\mu)^2}{2\sigma^2}\right]} \tag{5.55}$$

When $\mu = 0$ and $\sigma = 1$, the function that follows the Gaussian distribution is expressed by the equation (Melo and Watada, 2016):

$$g(x) = \frac{1}{\sqrt{2\pi}} e^{\left[\frac{-(x)^2}{2}\right]} \tag{5.56}$$

The Gaussian mutation operator was incorporated into the original bat algorithm to implement the mutation of the global best position according to the equation:

$$X_{best}^* = X_{best} + X_{best} Gauss(0,1) \tag{5.57}$$

where *Gauss(0,1)* represents a random number which follows the Gaussian distribution.

References

Boudjemaa, R., Oliva, D. and Ouaar, F. (2020). Fractional Lévy flight bat algorithm for global optimisation. *International Journal of Bio-Inspired Computation*, 15(2): 100–112.

Brown, C. T., Liebovitch, L. S. and Glendon, R. (2007). Lévy flights in Dobe Ju/'hoansi foraging patterns. *Human Ecology*, 35(1): 129–138.

Gandomi, A. H., Yang, X. S., Alavi, A. H. and Talatahari, S. (2013). Bat algorithm for constrained optimization tasks. *Neural Computing and Applications*, 22(6): 1239–1255.

Gandomi, A. H. and Yang, X.-S. (2014). Chaotic bat algorithm. *Journal of Computational Science*, 5(2): 224–232. doi: https://doi.org/10.1016/j.jocs.2013.10.002.

Gao, S., Gao, Y., Zhang, Y. and Xu, L. (2019). Multi-strategy adaptive cuckoo search algorithm. *IEEE Access*, 7: 137642–137655. doi:10.1109/ACCESS.2019.2916568.

Ghanem, W. A. H. M. and Jantan, A. (2019). An enhanced Bat algorithm with mutation operator for numerical optimization problems. *Neural Computing and Applications*, 31(1): 617–651. doi:10.1007/s00521-017-3021-9.

Gherboudj, A., Layeb, A. and Chikhi, S. (2012). Solving 0-1 knapsack problems by a discrete binary version of cuckoo search algorithm. *International Journal of Bio-Inspired Computation*, 4(4): 229–236.

Grieves, L. A. and Quinn, J. S. (2018). Group size, but not manipulated whole-clutch egg color, contributes to ovicide in joint-nesting Smooth-billed Anis. *The Wilson Journal of Ornithology*, 130(2): 479–484.

Hauber, M. E., Dainson, M., Baldassarre, D. T., Hossain, M., Holford, M. and Riehl, C. (2018). The perceptual and chemical bases of egg discrimination in communally nesting greater anis Crotophaga major. *Journal of Avian Biology*, 49(8): e01776.

Hughes, J. M. (1996). Phylogenetic analysis of the Cuculidae (Aves, Cuculiformes) using behavioral and ecological characters. *The Auk*, 113(1): 10–22.

Liao, C. J., Tseng, C. T. and Luarn, P. (2007). A discrete version of particle swarm optimization for flowshop scheduling problems. *Computers & Operations Research*, 34(10): 3099–3111.

Liu, Q., Li, J., Wu, L., Wang, F. and Xiao, W. (2020). A novel bat algorithm with double mutation operators and its application to low-velocity impact localization problem. *Engineering Applications of Artificial Intelligence*, 90: 103505. doi:https://doi.org/10.1016/j.engappai.2020.103505.

Luo, Q., Zhou, Y., Xie, J., Ma, M. and Li, L. (2014). Discrete bat algorithm for optimal problem of permutation flow shop scheduling. *The Scientific World Journal*, *2014*.

Lyu, S., Li, Z., Huang, Y., Wang, J. and Hu, J. (2019). Improved self-adaptive bat algorithm with step-control and mutation mechanisms. *Journal of Computational Science*, 30: 65–78. doi:https://doi.org/10.1016/j.jocs.2018.11.002.

Macedo, R. H. (1992). Reproductive patterns and social organization of the communal Guira Cuckoo (Guira guira) in central Brazil. *The Auk*, 109(4): 786–799.

Melo, H. and Watada, J. (2016). Gaussian-PSO with fuzzy reasoning based on structural learning for training a Neural Network. *Neurocomputing*, 172: 405–412.

Mirjalili, S., Mirjalili, S. M. and Yang, X.-S. (2014). Binary bat algorithm. *Neural Computing and Applications*, 25(3): 663–681. doi:10.1007/s00521-013-1525-5.

Mlakar, U., Fister, I. and Fister, I. (2016). Hybrid self-adaptive cuckoo search for global optimization. *Swarm and Evolutionary Computation*, 29: 47–72. doi:https://doi.org/10.1016/j.swevo.2016.03.001.

Nakamura, R. Y. M., Pereira, L. A. M., Costa, K. A., Rodrigues, D., Papa, J. P. and Yang, X. S. (2012, 22–25 Aug. 2012). *BBA: A Binary Bat Algorithm for Feature Selection.* Paper presented at the 2012 25th SIBGRAPI Conference on Graphics, Patterns and Images.

Paiva, F. A., Silva, C. R., Leite, I. V., Marcone, M. H. and Costa, J. A. (2017, November). Modified bat algorithm with cauchy mutation and elite opposition-based learning. pp. 1–6. *In: 2017 IEEE Latin American Conference on Computational Intelligence (LA-CCI).* IEEE.

Pavlyukevich, I. (2007). Cooling down Lévy flights. *Journal of Physics A: Mathematical and Theoretical*, 40(41): 12299.

Payne, R. B. and Sorensen, M. D. (2005). *The Cuckoos* (Vol. 15). Oxford University Press.

Pongchairerks, P. (2009). Particle swarm optimization algorithm applied to scheduling problems. *Science Asia*, 35: 89–94. doi: 10.2306/scienceasia1513-1874.2009.35.089.

Reynolds, A. M. and Frye, M. A. (2007). Free-flight odor tracking in Drosophila is consistent with an optimal intermittent scale-free search. *PloS One*, 2(4): e354.

Sharma, H., Bansal, J. C. and Arya, K. V. (2013). Opposition based lévy flight artificial bee colony. *Memetic Computing*, 5(3): 213–227.

Storn, R. and Price, K. (1997). Differential evolution—a simple and efficient heuristic for global optimization over continuous spaces. *Journal of Global Optimization*, 11(4): 341–359.

Wang, G.-G., Deb, S., Gandomi, A. H., Zhang, Z. and Alavi, A. H. (2016). Chaotic cuckoo search. *Soft Computing*, 20(9): 3349–3362. doi:10.1007/s00500-015-1726-1.

Yang, D., Li, G. and Cheng, G. (2007). On the efficiency of chaos optimization algorithms for global optimization. *Chaos, Solitons & Fractals*, 34(4): 1366–1375.

Yang, X. J. and Huang, Z. G. (2012). Opposition-based artificial bee colony with dynamic cauchy mutation for function optimization. *International Journal of Advancements in Computing Technology*, 4(4): 56–62.

Yang, X. S. and Deb, S. (2009, December). Cuckoo search via Lévy flights. pp. 210–214. *In: 2009 World Congress on Nature & Biologically Inspired Computing (NaBIC).* IEEE.

Yang, X. S. (2010a). A new metaheuristic bat-inspired algorithm. pp. 65–74. *In: Nature Inspired Cooperative Strategies for Optimization (NICSO 2010).* Springer, Berlin, Heidelberg.

Yang, X. S. (2010b). *Nature-inspired Metaheuristic Algorithms.* Luniver Press.

Yang, X. S. and Deb, S. (2010). Engineering optimisation by cuckoo search. *International Journal of Mathematical Modelling and Numerical Optimisation*, 1(4): 330–343.

Zhou, Y., Ling, Y. and Luo, Q. (2018). Lévy flight trajectory-based whale optimization algorithm for engineering optimization. *Engineering Computations*.

CHAPTER 6

Firefly Algorithm, Harmony Search and Cat Swarm Algorithm

◇◇◇

This chapter is devoted to the swarm optimization algorithms: Firefly algorithm, Harmony Search and Cat Swarm Optimization algorithm. The standard swarm optimization algorithms and some significant variations are discussed such as: Firefly algorithm with Lévy Flights Chaotic Firefly algorithm, Improved Harmony Search, Chaotic Harmony Search and Binary Discreate Cat Swarm Optimization and Improved Cat Swarm Optimization. Also, some important variations of the Bat Algorithm are discussed such as: Chaotic Bat Algorithm, Discreate Bat Algorithm and Binary Bat Algorithm.

Firefly algorithm

The firefly algorithm was developed by idealizing the flashing characteristics of natural fireflies in order to develop a computational optimization algorithm (Yang, 2009b, 2010b). The following rules were used to idealize the natural behavior of fireflies:

- All the fireflies are unisex so that one firefly will be attracted to other fireflies regardless of sex.
- The attractiveness is proportional to their brightness (light intensity), so in any pair of flashing fireflies, the lesser bright one will be attracted by the brighter one. When the distance increases the attractiveness

between them decreases. If a firefly cannot find a brighter one then it will move in a random way.

- The firefly brightness depends on the value of the objective function.

For simplicity reasons, it is assumed that the firefly attractiveness is defined by its brightness which is associated with the fitness function. The attractiveness β will be dependant on the distance r_{ij} between the firefly i and the firefly j. The firefly light intensity $I(r)$ decreases according to the distance from its source, and also the light varies according to the degree of absorption.

In the firefly algorithm, there are four important issues: the light intensity, the attractiveness, the firefly distance and the firefly movement.

Light intensity: In the firefly algorithm as proposed by Yang (2010), the firefly light intensity $I(r)$ which expresses a solution, for a given light absorption coefficient γ, varies with the distance r according to the equation:

$$I(r) = I_0 e^{-\gamma r^2} \tag{6.1}$$

where,

- I_0 represents the original light intensity
- r represents the distance between two fireflies
- γ represents a constant light absorption coefficient.

Attractiveness: The firefly attractiveness $\beta(r)$ is proportional to the light intensity and is expressed by the following equation:

$$\beta(r) = \beta_0 e^{-\gamma r^2} \tag{6.2}$$

where β_0 represents the attractiveness when the distance r = 0.

Distance: The distance $r_{i,j}$ between two fireflies i and j at X_i and X_j locations can be estimated according to the standard equation of the Cartesian distance:

$$r_{i,j} = \| X_i - X_j \| = \sqrt{\sum_{k=1}^{d} (x_{i,k} - x_{j,k})^2} \tag{6.3}$$

where,

- $x_{i,k}$ represents the k-th element of the i-th firefly
- d represents the dimension of the objective function that has to be optimized.

112

Movement: The movement of an i firefly attracted to another brighter - more attractive firefly j, can be calculated by the following equation:

$$X_i = X_i + \beta_0 e^{-\gamma r_{i,j}^2} \left(X_j - X_i \right) + \alpha \varepsilon_i \qquad (6.4)$$

where,

- α represents a randomized parameter
- ε_i represents a vector of random numbers with a Gaussian or uniform distribution.

According to Yang (2010), the simplest form of ε_i is expressed by:

$$\varepsilon_i = random - \frac{1}{2} \qquad (6.5)$$

where *random* is a randomly generated number uniformly distributed in the range [0,1]. So, the movement of an i firefly in this case can be expressed as:

$$X_i = X_i + \beta_0 e^{-\gamma r_{i,j}^2} \left(X_j - X_i \right) + \alpha \left(rand - \frac{1}{2} \right) \qquad (6.6)$$

The Firefly Algorithm is presented below (Yang, 2009, 2010b).

Firefly Algorithm

```
begin

    Objective function f(X); with X = (x₁,…,xₐ)ᵀ

    Generate the initial firefly population Xᵢ (i = 1,…,n)

    Light intensity Iᵢ at Xᵢ is defined by f(Xᵢ)

    Define the light absorption coefficient gamma(γ)

    while (t < MaximumGeneration)

        for i = 1:n to all n fireflies do

            for j = 1: n to all n fireflies do

                if (Iᵢ < Iⱼ) then

                    Move the firefly i to j;

                end if
```

```
        Attractiveness varies with distance r via e⁻ʸʳ²

        Evaluate new solutions and update light intensity

    end for

  end for

    Rank the fireflies and discover the current global best

  end while

    Post-process the results and visualize them

end
```

Firefly algorithm variants

Several modifications of the ·Firefly algorithm were proposed in the literature such as, Firefly algorithm with Lévy Flights (Yang, 2010b), chaotic Firefly algorithm (Coelho et al., 2011; Gandomi et al., 2013), Firefly algorithm with neighborhood attraction (Wang et al., 2017), Firefly algorithm for clustering (Senthilnath et al., 2011)

Firefly algorithm with Lévy flights

Yang (2010b), proposed Firefly algorithm with Lévy Flights. The pseudocode of the proposed algorithm is described below:

Firefly Algorithm

```
begin

    Objective function f(X); with X = (x₁,…,xₐ)ᵀ

    Generate the initial firefly population Xᵢ (i = 1,…,n)

    Light intensity Iᵢ at Xᵢ is defined by f(Xᵢ)

    Define the light absorption coefficient gamma(γ)

    while (t <MaximumGeneration)

      for i = 1:n to all n fireflies do
```

```
    for j = 1: n to all n fireflies do
        if (Iᵢ < Iⱼ) then
            Move the firefly i to j via Lévy Flights;
        end if
        Attractiveness varies with distance r via e⁻ᵞʳ²
        Evaluate new solutions and update light intensity
    end for
end for
    Rank the fireflies and discover the current global best
end while
    Post-process the results and visualize them
end
```

The movement of a firefly i with Lévy Flights which is attracted to a brighter firefly j is defined by the equation:

$$X_i = X_i + \beta_0 e^{-\gamma r_{i,j}^2} \left(X_j - X_i \right) + \alpha \, sign\left(rand - \frac{1}{2} \right) \oplus L\acute{e}vy \qquad (6.7)$$

- the third term of the equation expresses is randomization parameter via Levy flights.
- the product $\oplus$ expresses the entry-wise multiplication operator.
- The $sign\left(rand - \dfrac{1}{2} \right)$ expresses a random sign or direction and the random step length is calculated by a Levy distribution. Where $rand \in [0,1]$.

In the firefly algorithm with Lévy Flights, the firefly motion is a random walk process performed to prevent the algorithm from sticking into a local optimum, where the step length of the search process is improved and the deviation is expressed by the equation:

$$Levy \sim u = t^{-\lambda}, 0 < \lambda < 3 \tag{6.8}$$

where,

- t represents a random parameter in the range of $(0, 1]$,
- λ represents a stability index.

Chaotic firefly algorithms

Coelho et al. (2011), proposed a chaotic firefly algorithm by using Logistic map chaotic sequences in order to tune the parameters, α and γ. Consequently, the basic equation of the firefly algorithm altered to:

$$x_i = x_i + \beta_0 e^{-\gamma(t) \cdot r_{ij}^2} (x_j - x_i) + \alpha(t) \left(rand - \frac{1}{2} \right) \tag{6.9}$$

- $\gamma(t)$ is calculated by the equation:

$$\gamma(t) = \mu_1 \cdot \gamma(t-1) \cdot [1 - \gamma(t-1)] \tag{6.10}$$

- $\alpha(t)$ is calculated by the equation:

$$\alpha(t) = \mu_2 \cdot \gamma(t-1) \cdot [1 - \alpha(t-1)] \tag{6.11}$$

where,

- t represents the sample,
- μ_1 and μ_2 represent the control parameters, with $0 \leq \mu_1$ and $\mu_2 \leq 4$.

The authors referred that the proposed chaotic firefly algorithm outperformed the original firefly algorithm.

Yang (2013), proposed a chaos-enhanced firefly algorithm by also using automatic parameter tuning. In this research, a logistic map was used as a chaotic parameter for the attractiveness and the absorption coefficient replacing the Gaussian and the Lévy flight randomly distributed parameters. The proposed algorithm was an applied function optimization.

Arul et al. (2013), proposed a chaotic firefly algorithm to solve economic load dispatch problems by using the chaotic tent map in order to produce new values for the parameters, α and β_0. The equation for the Tent map is as follows:

$$\alpha_{n+1} = \begin{cases} 2\alpha_n & \alpha_n < 0.5 \\ 2(1-\alpha_n) & \alpha_n \geq 0.5 \end{cases} \qquad (6.12)$$

Harmony search algorithm

Harmony Search was proposed as a heuristic optimization algorithm (Geem, 2000; Geem et al., 2001). It was inspired by music harmony which is a combination of individual sounds aesthetically pleasant to the ear. It mimics a music orchestra while composing music in order to achieve the most harmonious melody. The procedure of searching for better harmony can be analogous to the process of finding an optimal solution to a computational optimization problem. Harmony Search has been applied on many kinds of optimization problems such as water and groundwater management, economic and emission dispatch, engineering, robotics, vehicle routing, structural design, energy, stock price prediction, medical diagnosis, and others (Ayvaz, 2009; Ceylan and Ceylan, 2009; Chatterjee et al., 2012; Guha et al., 2020; Jiang et al., 2020; Kayabekir et al., 2021; Mousavi et al., 2021).

As a musical orchestra tries to find the best state of harmony (fantastic harmony) which is defined aesthetically, an optimization algorithm searches for a global optimum by evaluating an objective function. As an aesthetic evaluation is defined by the set of sounds played by the individual instruments of the orchestra, an objective function evaluation is defined by the set of values generated. Each practice of the orchestra for a better sound is analogous to each iteration of an optimization algorithm in order to improve the objective function evaluation (Geem et al., 2001).

The main steps of the initially proposed algorithm are as follows (Geem et al., 2001)

1st Step. Initialize a Harmony Memory.

2nd Step. Create a new harmony from Harmony Memory.

3rd Step. If the new created harmony (solution) is improved compared to the minimum harmony in the Harmony Memory, then include it in the Harmony Memory, and remove the previous minimum harmony from it.

4th Step. If the predefined stopping criteria are not met, go to Step 2.

The Harmony Search holds the history of past vectors (solutions) (Harmony Memory) and can alter the adaptation rate by using the harmony memory accepting rate.

The harmony memory (HM) is used to store the best harmonies of new solution vectors. The harmony memory acceptance (or considering) rate ($r_{accept} \in [0,1]$) is used to effectively manipulate it. The pitch adjustment is accomplished by using pitch bandwidth b_{range} and also the pitch adjusting rate r_{pitch}. The pitch can be adjusted according to the following equation:

$$x_{new} = x_{old} + b_{range} \times \varepsilon \tag{6.13}$$

where,

- x_{old} represents the existing pitch in the harmony memory,
- x_{new} represents the new pitch after the pitch adjustment.
- b_{range} is the pitch bandwidth distance (Also, expressed as bw in the literature).
- ε represents a random number uniformly distributed in the range [−1, 1].

The pitch adjusting rate r_{pitch} controls the degree of adjustment. The most common values of r_{pitch} are in the range 0.1 ~ 0.5 and r_{accept} in the range 0.7 ~ 0.95 (Yang, 2009a). According to Yang (2009a) the harmony search is described by the following algorithm:

Harmony search algorithm

```
begin
Determine the objective function f(x), where x=(x₁,x₂, …,xₐ)ᵀ
Determine the harmony memory accepting rate r_accept
Determine the pitch adjusting rate (r_pitch) and the other parameters
Generate Harmony Memory with random harmonies
while (t<max number of iterations)
      while (i<=NumberOfVariables)
         if (rand<r_acc) then
```

```
        Choose a value from HM for the variable i
        if (rand<r_pitch) then
            Adjust the value by adding certain amount
        end if
        else
            Choose a random value
        end if
    end while
    Accept the new solution if it is better
  end while
  Find the current best solution
end
```

Harmony search example

Ackley function approximation

Mathematical equation

The equation that represents the Ackley function is as follows:

$$f(x_1,...,x_n) = -a \cdot exp(-b\sqrt{\frac{1}{n}\sum_{i=1}^{n}x_i^2}) - exp(\frac{1}{n}\sum_{i=1}^{n}cos(cx_i)) + a + exp(1) \quad (6.14)$$

where a, b and c are constants. For the experiments we set: a = 20, b = 0.2 and c = 2π. The Ackley function is continuous, non-convex, multimodal and differentiable.

Input search space

The function is evaluated with $x_i \in [-20,20]$, and i = 1, 2, . . . , n.

Harmony Search Parameters

Maximum Number of Iterations: 10000

Harmony Memory Size: 25

Number of New Harmonies: 20

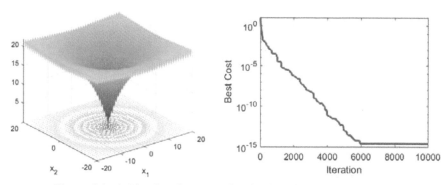

Figure 6.1. Ackley function approximation by using Harmony Search.

Harmony Memory Accepting Rate: $r_{accept} = 0.9$

Pitch Adjustment Rate: $r_{pitch} = 0.1$

Decision Variables Minimum Bound (DVMinB): –20

Decision Variables Maximum Bound (DVMaxB): 20

Pitch Bandwidth:

$$b = 0.02 * (DVMaxB - DVMinB) \qquad (6.15)$$

The optimization results illustrated in the next figure are very good. The function minimization stops improving at about 6000 iterations.

Harmony search variants

Improved harmony search algorithm

Several researches have tried to improve the convergence of the Harmony Search algorithm. Mahdavi et al. (2007), proposed an improved harmony search algorithm. According to the authors, in the improved harmony search algorithm the pitch adjusting rate *PAR* (Or r_{pitch}) is dynamically updated by using the following equation:

$$PAR(t) = PAR_{\min} + \frac{(PAR_{\max} - PAR_{\min})}{NI} \times t \qquad (6.16)$$

where,

- *PAR(t)* represents the pitch adjusting rate for the iteration *t*.
- $PAR_{\min}$ represents the minimum adjusting rate.

- PAR_{max} represents the maximum adjusting rate.
- NI represents the number of the iteration.
- t represents the iteration number.

Also, in the improved harmony search algorithm, the bandwidth distance parameter bw (Or b_{range}) is dynamically updated by using the following equation:

$$bw(t) = bw_{max} e^{\left(\frac{\ln\left(\frac{bw_{min}}{bw_{max}}\right)}{NI} \times t \right)}$$

(6.17)

where,

- $bw(t)$ represents the bandwidth distance for iteration t.
- bw_{min} represents the minimum bandwidth.
- bw_{max} represents the maximum bandwidth.

Four new parameters are introduced in the improved harmony search algorithm: PAR_{min}, PAR_{max}, bw_{min}, and bw_{max}.

Chaotic harmony search

In the Harmony Search algorithm, there are parameters with constant values during the algorithm process. Alatas (2010), proposed some methods to apply chaotic sequences in order to substitute the random values of the parameters in order to improve the global convergence and prevent the algorithm from being trapped in a local minimum.

In the proposed algorithm, pitch adjusting rate PAR (Or r_{pitch}) value is not constant in the parameter initialization stage and it has been substituted by chaotic maps as in the following equation:

$$PAR(t+1) = f(PAR(t)), \quad 0 < PAR(t) < 1, \quad t = 0,1,2,\ldots \quad (6.18)$$

where the function f is a chaotic map.

Also, Alatas (2010), proposed that bandwidth distance b_w (Or b_{range}) value can be substituted in the parameter initialization stage by chaotic maps according to the following equation:

$$bw(t+1) = f(bw(t)), \quad 0 < bw(t) < 1, \quad t = 0,1,2,\ldots \quad (6.19)$$

where the function f is a chaotic map.

The authors have used seven chaotic maps for the experiments:

Logistic map, expressed by the equation:

$$X_{n+1} = aX_n(1-X_n) \tag{6.20}$$

Tent map, expressed by the equations:

$$X_{n+1} = \begin{cases} X_n/0.7, & X_n < 0.7, \\ 10/3 X_n(1-X_n), & \text{otherwise.} \end{cases} \tag{6.21}$$

Sinusoidal iterator, expressed by the equation:

$$X_{n+1} = ax_n^2 \sin(\pi x_n) \tag{6.22}$$

with a = 2.3 and $X_0 = 0.7$

Gauss map is expressed by the equations:

$$X_{n+1} = \begin{cases} 0, & X_n = 0, \\ 1/X_n \bmod (1), & X_n \in (0,1), \end{cases} \tag{6.23}$$

$$1/X_n \bmod(1) = \frac{1}{X_n} - \frac{1}{X_n} \tag{6.24}$$

Circle map is expressed by the equation:

$$X_{n+1} = X_n + b - (a/2\pi)\sin(2\pi X_n) \bmod (1) \tag{6.25}$$

with a = 0.5 and b = 0.2.

Sinus map is expressed by the equation:

$$X_{n+1} = 2.3(X_n)^{2\sin(\pi X_n)} \tag{6.26}$$

Henon map is expressed by the equations:

$$X_{n+1} = 1 - aX_n^2 + bY_n \tag{6.27}$$

$$Y_{n+1} = X_n \tag{6.28}$$

$$X_{n+1} = 1 - aX_n^2 + bX_{n-1} \tag{6.29}$$

with parameter values a = 1.4 and b = 0.3.

Cat swarm optimization

Cat swarm algorithm

The Cat Swarm Algorithm was introduced by Chu et al. (2006). The authors studied the behavior of cats in order to introduce a new optimization algorithm. They have proposed that cats have two modes: the seeking mode and the tracing mode. They observed that cats spend most of their time resting and observing. In their resting and observing time, they move slowly and carefully. This can be called the seeking mode. Cats spend little time in chasing things. This is called tracing mode, where cats move according to their own velocities (Chu et al., 2006; Chu and Tsai, 2007).

Seeking mode

Seeking mode has four basic parameters: the seeking memory pool (SMP), the seeking range of the selected dimension (SRD), the counts of dimension to change (CDC), and the self-position considering (SPC).

- **SMP** is utilized to determine the size of the seeking memory for each cat.
- **SRD** represents the mutation ratio for the selected dimensions.
- **CDC** represents how many dimensions will be changed.
- **SPC** represents a Boolean parameter, which is used as a decision variable to define if the current cat point will be one of the candidate points (solutions) to move to.

The seeking mode consists of five simple steps (Chu et al. 2006):

Step 1: Make j copies of the current position of cat_k, with $j = SMP$. If the SPC value is true (Equal to one), $j = (SMP - 1)$, then retain the current position as one of the candidate solutions.

Step 2: For every copy, according to the counts of dimension to change (CDC), randomly add or subtract the selected dimension SRD percentages from the current values and replace the old values.

Step 3: Estimate the fitness values of all candidate solutions.

Step 4: If all fitness function values are not the same, estimate the selection probability of every candidate by using the following equation, otherwise set each candidate selection probability to 1.

$$P_i = \frac{FS_i - FS_b}{FS_{max} - FS_{min}} \tag{6.30}$$

where,

- FS_i is the fitness of the *ith* cat
- $FS_b = FS_{max}$ if the minimum solution should be found
- $FS_b = FS_{min}$ if the maximum solution should be found.

Step 5: Pick in a random way the point to move to from the candidate position, and replace the current position of cat$_k$ with the selected candidate.

Tracing mode

In tracing mode, the cat is tracing targets. In this mode, the next position of every cat is defined by the individual cat velocity and the cat swarm best position. This mode can be summarized in the following steps.

Step 1: Update the cat velocities for every dimension ($v_{k,d}$) according to the equation:

$$v_{k,d} = v_{k,d} + r_1 c_1 \left(x_{gbest,d} - x_{k,d} \right), \text{ with } d = 1, \ldots, M \tag{6.31}$$

where,

- $x_{gbest,d}$ represents the position of the cat with the best fitness value
- $x_{k,d}$ represents the current position of the cat, in d*th* dimension
- c_1 represents a fix value in the interval [0, 2]
- r_1 represents a uniformly distributed random value in the space [0, 1].

Step 2: Check if the velocities are within the range bounds of the maximum velocity. If the new velocity is out of the range, then define it as equal to the maximum velocity.

Step 3: Update the current position of cat$_k$ using the following equation:

$$x_{k,d} = x_{k,d} + v_{k,d} \tag{6.32}$$

Basic description of Cat Swarm Algorithm

The Cat Swarm Algorithm has two modes, the seeking mode and the tracing mode. In order to combine these two processes, a mixture ratio is defined which sets the rate of cats in seeking mode and the rest of the cats in tracing mode. According to Chu et al. (2006), the Cat Swarm Algorithm consists of the following stages:

Stage 1: Create N cats.

Stage 2: Scatter the cats into a M-dimensional solution space randomly and set random values, within the range of the maximum velocity for every cat. Then randomly pick a number of cats and set them into tracing process according to the mixture ratio and other cats into seeking process.

Stage 3: Evaluate the fitness value of every cat by applying the positions (solutions) x_i of cats into the fitness function, and keep the best cat into memory (The best solution so far).

Stage 4: Move the cats according to their categorization. If a cat is in seeking mode, follow the seeking mode procedure, otherwise follow the tracing mode procedure.

Stage 5: Pick again a number of cats and put them into tracing mode according to the mixture ratio and the others into seeking mode.

Stage 6: Check the termination criterion, if it is met, terminate, otherwise repeat Step 3 to Step 5.

Cat algorithm variants

Binary discrete Cat algorithm

The original Cat Swarm Algorithm was developed for continuous optimization problems. Sharafi et al. (2013), proposed a binary discrete cat algorithm. In the proposed binary discrete cat algorithm, the cat position vector consists of ones and zeros. In the Seeking Mode of Binary CSO, all of the values are zero and one the current position of the cat can be determined as a binary mutation. In Binary CSO the parameter probability of the mutation operation (PMO) substitutes the parameter SRD of the original CSO. In the tracing mode of Binary CSO, the main difference is in the velocity definition. The velocity vector in Binary CSO represents the probability of change and is updated as follows. The two velocity vectors, one for every cat, are determined as V_{kd}^1 and V_{kd}^0. V_{kd}^1 representing the probability of the bits to invert from one to zero and V_{kd}^0 represents the probability of the bits to invert from zero to one.

The velocities V^1_{kd} and V^0_{kd} are calculated by the following equations:

$$V^1_{kd} = wV^1_{kd} + d^1_{kd} \tag{6.33}$$

$$V^0_{kd} = wV^0_{kd} + d^0_{kd} \qquad d = 1,...,M \tag{6.34}$$

d^1_{kd} and d^0_{kd} are updated according to the following equations:

$$\text{if } X_{gbest,d} = 1 \text{ then } d^1_{kd} = r_1 c_1 \text{ and } d^0_{kd} = -r_1 c_1 \tag{6.35}$$

$$\text{if } X_{gbest,d} = 0 \text{ then } d^1_{kd} = -r_1 c_1 \text{ and } d^0_{kd} = r_1 c_1 \tag{6.36}$$

where,

- r_1 represents random value in the interval $[0, 1]$.
- w represents the inertia weight.
- c_1 represents a fixed value determined by the user.

The velocity of the cat$_k$ is estimated by the following equation:

$$V'_{kd} = \begin{cases} V^1_{kd} & \text{if } X_{kd}=0 \\ V^0_{kd} & \text{if } X_{kd}=1 \end{cases} \tag{6.37}$$

The mutation probability in every dimension is determined by the variable t which is estimated by the following equation.

$$t_{kd} = sig(V'_{kd}) = \frac{1}{1+e^{-V'_{kd}}} \tag{6.38}$$

where t_{kd} takes values in the range $[0, 1]$. The new cat position of every dimension is updated according to the following equation.

$$x_{kd} = \begin{cases} X_{gbest,d} & \text{if } rand < t_{kd} \\ x_{kd} & \text{if } t_{kd} < rand \end{cases} \quad d = 1,...M \tag{6.39}$$

If the value of V'_{kd} has a value larger than the maximum velocity (out of maximum bound), then maximum velocity is selected as the velocity value.

Improved cat swarm optimization

Kumar and Singh (2018), proposed an improved cat swarm algorithm in order to solve global optimization problems. The CSO algorithm suffers from several problems, such as: premature convergence (Orouskhani et al., 2011, 2013), it cannot explore all good solutions because of the lack of an information mechanism about the global best cat position.

In order to improve the CSO algorithm, the following modifications are proposed by Kumar and Singh (2018).

The global best cat position is used in order to explore better solutions and improve the convergence rate and lead the cats in the tracing mode. The new proposed modified equation in tracing mode of the algorithm includes the global best cat position:

$$X_{j\,new}^{d+1} = (1-\beta) * X_j^d + \beta * P_g + V_j^d \qquad (6.40)$$

The authors, proposed a new velocity update equation, in order to improve the cat swarm optimization algorithm diversity, in tracing mode:

$$V_{j\,new}^{d+1} = V_j^d + \beta\left(P_g - X_j^d\right) + \alpha * \varepsilon \qquad (6.41)$$

where,

- ε represents a vector randomly and uniformly distributed in the space [0, 1].
- α and β represent the acceleration parameters for directing the cat position into a local and global best.
- P_g represents the global best position of a cat.

The values of both α and β acceleration parameters for directing the cat position in a local and global best are adaptive and calculated by the following equations:

$$\alpha(t) = \alpha_{max} - \left\{\frac{\alpha_{max} - \alpha_{min}}{t_{max}}\right\} * t \qquad (6.42)$$

where,

- α_{max} and α_{min} represent the upper and lower limits.
- t_{max} represents the maximum number of the iterations.
- t represents the current number of the iteration.

$$\beta(t) = \beta_{\min} + (\beta_{\max} - \beta_{\min}) \sin\left\{\frac{\pi t}{t_{\max}}\right\} \tag{6.43}$$

where,

- $\beta_{\min}$ and $\beta_{\max}$ represent the minimum and maximum values of the first and the last iterations respectively.
- t_{max} represents the maximum number of iterations.
- t represents the current number of the iteration.

References

Alatas, B. (2010). Chaotic harmony search algorithms. *Applied Mathematics and Computation,* 216(9): 2687–2699. doi:https://doi.org/10.1016/j.amc.2010.03.114.

Arul, R., Velusami, S. and Ravi, G. (2013, December). Chaotic firefly algorithm to solve economic load dispatch problems. pp. 458–464. In: *2013 International Conference on Green Computing, Communication and Conservation of Energy (ICGCE).* IEEE.

Ayvaz, M. T. (2009). Identification of groundwater parameter structure using harmony search algorithm. pp. 129–140. In: *Music-inspired Harmony Search Algorithm.* Springer, Berlin, Heidelberg.

Ceylan, H. and Ceylan, H. (2009). Harmony search algorithm for transport energy demand modeling. *Music-Inspired Harmony Search Algorithm,* 163–172.

Chatterjee, A., Ghoshal, S. P. and Mukherjee, V. (2012). Solution of combined economic and emission dispatch problems of power systems by an opposition-based harmony search algorithm. *International Journal of Electrical Power & Energy Systems,* 39(1): 9–20.

Chu, S. C., Tsai, P. W. and Pan, J. S. (2006, August). Cat swarm optimization. pp. 854–858. In: *Pacific Rim International Conference on Artificial Intelligence.* Springer, Berlin, Heidelberg.

Chu, S. C. and Tsai, P. W. (2007). Computational intelligence based on the behavior of cats. *International Journal of Innovative Computing, Information and Control,* 3(1): 163–173.

Coelho, L. d. S., Bernert, D. L. d. A. and Mariani, V. C. (2011, 5–8 June 2011). *A Chaotic Firefly Algorithm Applied to Reliability-Redundancy Optimization.* Paper presented at the 2011 IEEE Congress of Evolutionary Computation (CEC).

Gandomi, A. H., Yang, X. S., Talatahari, S. and Alavi, A. H. (2013). Firefly algorithm with chaos. *Communications in Nonlinear Science and Numerical Simulation,* 18(1): 89–98. doi:https://doi.org/10.1016/j.cnsns.2012.06.009.

Geem, Z. W. (2000). Optimal Design of Water Distribution Networks using Harmony Search. Ph.D. Thesis, Korea University.

Geem, Z. W., Kim, J. H. and Loganathan, G. V. (2001). A new heuristic optimization algorithm: harmony search. *Simulation*, 76(2): 60–68.

Guha, S., Das, A., Singh, P. K., Ahmadian, A., Senu, N. and Sarkar, R. (2020). Hybrid feature selection method based on harmony search and naked mole-rat algorithms for spoken language identification from audio signals. *IEEE Access*, 8: 182868–182887.

Jiang, M., Jia, L., Chen, Z. and Chen, W. (2020). The two-stage machine learning ensemble models for stock price prediction by combining mode decomposition, extreme learning machine and improved harmony search algorithm. *Annals of Operations Research*, 1–33.

Kayabekir, A. E., Bekdas, G., Yücel, M., Nigdeli, S. M. and Geem, Z. W. (2021). Harmony search algorithm for structural engineering problems. *Nature-Inspired Metaheuristic Algorithms for Engineering Optimization Applications*, 13.

Kumar, Y. and Singh, P. K. (2018). Improved cat swarm optimization algorithm for solving global optimization problems and its application to clustering. *Applied Intelligence*, 48(9): 2681–2697.

Mahdavi, M., Fesanghary, M. and Damangir, E. (2007). An improved harmony search algorithm for solving optimization problems. *Applied Mathematics and Computation*, 188(2): 1567–1579.

Mousavi, S. M., Abdullah, S., Niaki, S. T. A. and Banihashemi, S. (2021). An intelligent hybrid classification algorithm integrating fuzzy rule-based extraction and harmony search optimization: Medical diagnosis applications. *Knowledge-based Systems*, 220: 106943.

Orouskhani, M., Mansouri, M. and Teshnehlab, M. (2011). Average-inertia weighted cat swarm optimization. pp. 321–328. *In*: *International Conference in Swarm Intelligence*. Springer, Berlin, Heidelberg.

Orouskhani, M., Orouskhani, Y., Mansouri, M. and Teshnehlab, M. (2013). A novel cat swarm optimization algorithm for unconstrained optimization problems. *International Journal of Information Technology and Computer Science*, 5(11): 32–41.

Senthilnath, J., Omkar, S. N. and Mani, V. (2011). Clustering using firefly algorithm: Performance study. *Swarm and Evolutionary Computation*, 1(3): 164–171. doi:https://doi.org/10.1016/j.swevo.2011.06.003.

Sharafi, Y., Khanesar, M. A. and Teshnehlab, M. (2013, 25–26 Sept. 2013). *Discrete Binary Cat Swarm Optimization Algorithm*. Paper presented at the 2013 3rd IEEE International Conference on Computer, Control and Communication (IC4).

Wang, H., Wang, W., Zhou, X., Sun, H., Zhao, J., Yu, X. and Cui, Z. (2017). Firefly algorithm with neighborhood attraction. *Information Sciences*, 382-383: 374–387. doi:https://doi.org/10.1016/j.ins.2016.12.024.

Yang, X. S. (2009a). Harmony search as a metaheuristic algorithm. pp. 1–14. *In*: *Music-inspired Harmony Search Algorithm*. Springer, Berlin, Heidelberg.

Yang, X. S. (2009b). Firefly algorithms for multimodal optimization. pp. 169–178. *In*: *International Symposium on Stochastic Algorithms*. Springer, Berlin, Heidelberg.

Yang, X. S. (2010a). *Nature-inspired Metaheuristic Algorithms*. Luniver Press.

Yang, X. S. (2010b). Firefly algorithm, Levy flights and global optimization. pp. 209–218. *In*: *Research and Development in Intelligent Systems XXVI*. Springer, London.

Yang, X. S. (2013). Chaos-enhanced firefly algorithm with automatic parameter tuning. pp. 125–136. *In*: *Recent Algorithms and Applications in Swarm Intelligence Research*. IGI Global.

CHAPTER 7

Grey Wolf, Whale and Grasshopper Optimization

This chapter is devoted to the latest bio-inspired swarm algorithms such as: Grey Wolf Optimization (GWO) Algorithm (2014), Whale Optimization Algorithm (WOA) (2016) and Grasshopper Optimization Algorithm (GOA) (2017). The original algorithms and some recent important variants are discussed such as: Binary Grey Wolf Optimization (2016), Grey Wolf with Lévy flight (2017), Whale Optimization with Lévy Flight (2017), Binary Whale Optimization Algorithm (2020), Improved Grasshopper Optimization Algorithm (2018) and Chaotic Grasshopper Optimization (2019).

Grey wolf optimization

The grey wolf optimizer was introduced by Mirjalili et al. (2014). The grey wolf swarm algorithm was inspired by the natural behavior of grey wolves in group hunting. The grey wolf optimizer is relatively simple compared to the swarm intelligence algorithm with fewer hyper-parameters to be set, compared to other optimization algorithms and also the implementation of the algorithm is relatively easy.

The grey wolf optimizer was applied on several kinds of optimization problems such as: economic dispatch (Jayabarathi et al., 2016; Wong et al., 2014), feature selection (Emary et al., 2015; Emary et al., 2016; Hu et al., 2020), engineering (Nadimi-Shahraki et al., 2021), forecasting electric loads (Zhang and Hong, 2021), wind speed forecasting (Altan

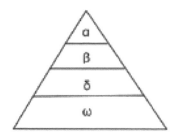

Figure 1. Grey wolves pack hierarchy.

et al., 2021), human activity recognition (Helmi et al., 2021), building energy optimization (Ghalambaz et al., 2021), energy consumption prediction and optimization (Li et al., 2021), power quality improvement (Thakur et al., 2021) and others.

Grey wolves are predators, and prefer to live in a pack. The social hierarchy of the Grey wolf pack is shown on the next figure. Alpha (α) is the highest level in the Grey wolf pack hierarchy.

Inspired by this pack hierarchy and the natural behavior of the grey wolf pack, Mirjalili et al. (2014) proposed the grey wolf optimizer where the population of grey wolves is separated into four groups: alpha, beta, delta and omega. The group separation is according to the wolf social hierarchy. Alpha represents the leader (The dominant wolf), the top wolf in the hierarchy, which represents the best solution. The second-best solution is represented by beta (β), third best solution is represented by delta (δ). The rest of the solutions are represented as omega (ω).

The hunting behavior of grey wolves consists of three main stages:

- searching for prey
- encircling the prey
- attacking the prey.

Encircling prey

The grey wolf pack is encircling the prey in the hunting process. Mirjalili et al. (2014) proposed the following mathematical equations to model the encircling behavior:

$$\vec{D} = | \vec{C} \cdot \vec{X}_p(t) - \vec{X}(t) | \qquad (7.1)$$

$$\vec{X}(t+1) = \vec{X}_p(t) - \vec{A} \cdot \vec{D} \qquad (7.2)$$

where,

- t represents the current iteration.
- $\vec{A}$ and $\vec{C}$ represent the coefficient vectors.
- $\vec{X}_p$ represents the prey position vector.
- $\vec{X}$ represents the grey wolf position vector.

The $\vec{A}$ and $\vec{C}$ vectors are estimated according to the following mathematical equations:

$$\vec{A} = 2\vec{a} \cdot \vec{r}_1 - \vec{a} \qquad (7.3)$$

$$\vec{C} = 2 \cdot \vec{r}_2 \qquad (7.4)$$

where,

- $\vec{a}$ is linearly decreased from 2 to 0 during the algorithm iterations
- r_1, r_2 represent random vectors in the interval [0, 1].

Hunting

Alpha (α), beta (β) and delta (δ) wolves participate in the hunting process and alpha (α) represents the best solution in the pack, beta (β) the second-best solution and delta (δ) the third best solution. The first three best solutions are saved and the other wolves update their current positions randomly in the search space. Mirjalili et al. (2014), proposed the following equations in order to simulate the Grey wolf hunting behavior.

$$\vec{D}_\alpha = |\vec{C}_1 \cdot \vec{X}_\alpha - \vec{X}| \qquad (7.5)$$

$$\vec{D}_\beta = |\vec{C}_2 \cdot \vec{X}_\beta - \vec{X}| \qquad (7.6)$$

$$\vec{D}_\delta = |\vec{C}_3 \cdot \vec{X}_\delta - \vec{X}| \qquad (7.7)$$

$$\vec{X}_1 = \vec{X}_\alpha - \vec{A}_1 \cdot (\vec{D}_\alpha) \qquad (7.8)$$

$$\vec{X}_2 = \vec{X}_\beta - \vec{A}_2 \cdot (\vec{D}_\beta) \qquad (7.9)$$

$$\vec{X}_3 = \vec{X}_\delta - \vec{A}_3 \cdot (\vec{D}_\delta) \qquad (7.10)$$

$$\vec{X}(t+1) = \frac{\vec{X}_1 + \vec{X}_2 + \vec{X}_3}{3} \qquad (7.11)$$

where,

- $\vec{X}_1, \vec{X}_2, \vec{X}_3$ represent the of alpha (α), beta (β) and delta (δ) wolf positions
- $\vec{X}(t+1)$ represents the updated next wolf position.
- $\vec{X}_\alpha, \vec{X}_\beta, \vec{X}_\gamma$ are the first three best solutions.
- $\vec{A}_1, \vec{A}_2, \vec{A}_3$ are calculated by using equation (7.3).
- $\vec{C}_1, \vec{C}_2, \vec{C}_3$ are defined by using equation (7.4).

Attacking prey (exploitation)

The hunting process of grey wolves finishes when the prey stops moving. Then the grey wolves start to attack. This is modelled by Mirjalili et al. (2014), by decreasing the value of $\vec{a}$ linearly during the iterations.

Search for prey (exploration)

Grey wolves search for solutions near the position of the alpha, beta, and delta. Mirjalili et al. (2014), use $\vec{A}$ with random values, bigger than 1 or less than -1 so as to force the grey wolves to diverge and enhance exploration in order to allow the algorithm to search globally in the search space.

The pseudocode of the Grey Wolf Optimization Algorithm as proposed by Mirjalili et al. (2014), is presented below.

Grey Wolf Optimization Algorithm

```
Initialize the population of grey wolves X_i (i = 1,2,…,n)

Initialize the parameters a, A and C

Estimate the fitness function value of each wolf agent

X_α = the best wolf search agent

X_β = the second-best wolf search agent

X_δ = the third best wolf search agent
```

```
while (t < MaximumNumberOfIterations) do

    for each search wolf do

        update the positions of current wolf agent using equation (7.11)

    end for

    update α, A, C

    Estimate the fitness function value for all wolf search agents

    update the values of X_α, X_β and X_δ

    t=t+1

end while

return the best wolf position X_α
```

Grey wolf algorithm variants

Binary grey wolf optimization

Emary et al. (2016), proposed the binary grey wolf optimization (bGWO) method for feature selection. The original grey wolf optimization is defined for continuous optimization problems in a continuous search space. In optimization problems with a binary nature the solutions of the algorithm are binary with values 0 and 1.

In the original grey wolf optimization algorithm, the wolves update their positions according to the first three best solutions $\vec{X}_\alpha, \vec{X}_\beta, \vec{X}_\gamma$. In the binary grey wolf optimization, the solutions are only binary.

Emary et al. (2016), proposed two binary grey wolf optimization methods.

Binary grey wolf optimization method I

In binary grey wolf optimization, the equation for the position update can be expressed as under,

$$X_i^{t+1} = Crossover\left(x_1, x_2, x_3\right) \tag{7.12}$$

where,

- *Crossover*(*x*, *y*, *z*) is the cross over operation among the solutions *x, y, z*

135

- x_1, x_2, x_3 represent binary vectors expressing the impact of the movement of a wolf in the direction of α, β, δ wolves, calculated by using the following equations:

$$x_1^d = \begin{cases} 1 & \text{if } \left(x_\alpha^d + bstep_\alpha^d\right) \geq 1 \\ 0 & \text{otherwise} \end{cases} \tag{7.13}$$

where,

- x_α^d represents the solution vector of the α wolf in dimension d,
- $bstep_\alpha^d$ represents a binary step in the d dimension calculated as by using the following equation:

$$bstep_\alpha^d = \begin{cases} 1 & \text{if } cstep_\alpha^d \geq rand \\ 0 & \text{otherwise} \end{cases} \tag{7.14}$$

where,

- *rand* represents a random number uniformly distributed in the range [0,1].
- $cstep_\alpha^d$ represents a continuous step value for the d dimension and can be estimated by using the following equation:

$$cstep_\alpha^d = \frac{1}{1 + e^{-10\left(A_1^d D_\alpha^d - 0.5\right)}} \tag{7.15}$$

where,

- A_1^d D_α^d are estimated by using the equations: (7.3) (7.5) in the d dimension.

$$x_2^d = \begin{cases} 1 & \text{if } \left(x_\beta^d + bstep_\beta^d\right) \geq 1 \\ 0 & \text{otherwise} \end{cases} \tag{7.16}$$

where,

- x_β^d represents the position (solution) vector of the β wolf in the d dimension,
- $bstep_\beta^d$ represents a binary step in the d dimension calculated by the equation:

$$bstep_\beta^d = \begin{cases} 1 & \text{if } cstep_\beta^d \geq rand \\ 0 & \text{otherwise} \end{cases} \tag{7.17}$$

where,

- *rand* represents a random number uniformly distributed in the range [0,1].
- $cstep_\beta^d$ represents a continuous step value for the d dimension and can be estimated by using the following equation:

$$cstep_\beta^d = \frac{1}{1 + e^{-10\left(A_1^d D_\beta^d - 0.5\right)}}$$ (7.18)

where,

- $A_1^d \, D_\beta^d$ are estimated by using the equations (7.3) and (7.5) in the d dimension.

$$x_3^d = \begin{cases} 1 & \text{if } \left(x_\delta^d + bstep_\delta^d\right) \geq 1 \\ 0 & otherwise \end{cases}$$ (7.19)

where,

- x_δ^d represents the position (solution) vector of the δ wolf in the d dimension.
- $bstep_\delta^d$ represents a binary step in the d dimension calculated by the following equation:

$$bstep_\delta^d = \begin{cases} 1 & \text{if } cstep_\delta^d \geq rand \\ 0 & otherwise \end{cases}$$ (7.20)

where,

- *rand* represents a random number uniformly distributed in the range [0,1].
- $cstep_\delta^d$ represents a continuous step value for the d dimension and can be estimated by using the following equation:

$$cstep_\delta^d = \frac{1}{1 + e^{-10\left(A_1^d D_\delta^d - 0.5\right)}}$$ (7.21)

where,

- $A_1^d \, D_\delta^d$ are estimated by using the equations: (7.3), (7.5) in the d dimension.

The crossover method is expressed by the following equation:

$$x_d = \begin{cases} a_d & \text{if } rand < \dfrac{1}{3} \\ b_d & \dfrac{1}{3} \leq rand < \dfrac{2}{3} \\ c_d & otherwise \end{cases} \tag{7.22}$$

where,

- a_d, b_d, c_d represent the binary values of the first, the second and the third variable in the d dimension
- x_d represents the output of the crossover
- *rand* represents a random number uniformly distributed in the range [0,1].

Binary grey wolf optimization method II

Emary et al. (2016), proposed a second binary grey wolf optimization method where the updated grey wolf position vector is converted to binary according to the equation:

$$x_d^{t+1} = \begin{cases} 1 & \text{if } sigmoid\left(\dfrac{x_1 + x_2 + x_3}{3}\right) \geq rand \\ 0 & otherwise \end{cases} \tag{7.23}$$

where,

- *rand* represents a random number uniformly distributed in the range [0,1].
- x_d^{t+1} represents the position (solution) vector of the next time step in the d dimension.
- t represents the iteration number.
- *sigmoid(a)* is calculated by the following equation:

$$sigmoid(a) = \frac{1}{1 + e^{-10(x - 0.5)}} \tag{7.24}$$

Grey wolf with Lévy flight

Heidari and Pahlavani (2017), proposed a modified grey wolf optimizer with Lévy flight to solve the immature convergence problem of the original algorithm. The authors proposed three modifications to the conventional GWO algorithm:

- the role of the δ wolves in the hierarchy can be played by other grey wolves.
- the Lévy flight is incorporated into the modified GWO algorithm.
- the greedy selection is applied on the proposed Lévy flight GWO.

In Lévy flight GWO, the hierarchy includes three types of wolves including alpha, beta, and omega. The position of the grey wolves in the Lévy flight GWO is estimated based on alpha and beta grey wolves by using the following equation:

$$\vec{X}(t) = 0.5 \times (\vec{X}_\alpha - \vec{A}_1\vec{D}_\alpha + \vec{X}_\beta - \vec{A}_2\vec{D}_\beta).$$ (7.25)

The Lévy function can be expressed by the equation:

$$L(s) \sim |s|^{-1-\beta}, 0 < \beta \le 2$$ (7.26)

where,

- s represents a variable.
- β represents the Lévy index for the stability.

The Lévy distribution can be expressed by the equation:

$$L(s,\gamma,\mu) = \begin{cases} \sqrt{\dfrac{\gamma}{2\pi}} \, exp\left[-\dfrac{\gamma}{2(s-\mu)}\right]\dfrac{1}{(s-\mu)^{3/2}} & 0 < \mu < s < \infty \\ 0 & s \le 0 \end{cases}$$ (7.27)

where,

- μ represents a shift parameter
- *gamma parameter* $\gamma > 0$ represents a scale parameter

Lévy distribution can also be expressed by using the Fourier transformation:

$$F(k) = exp\left[-\alpha|k|^\beta\right], 0 < \beta \le 2$$ (7.28)

where α represents the scale factor in the range [−1, 1].

139

Heidari and Pahlavani (2017), proposed another modification to the original GWO, regarding the new positions' equation:

$$\vec{X}_{new}(t) = \begin{cases} 0.5 \times (\vec{X}_\alpha - \vec{A}_1\vec{D}_\alpha + \vec{X}_\beta - \vec{A}_2\vec{D}_\beta) + \alpha \oplus Levi(\beta) & |A| > 0.5 \\ 0.5 \times (\vec{X}_\alpha - \vec{A}_1\vec{D}_\alpha + \vec{X}_\beta - \vec{A}_2\vec{D}_\beta) & |A| < 0.5 \end{cases} \quad (7.29)$$

where,

- α represents the step size.
- β represents the Lévy index inside [0, 2].
- $\oplus$ represents the entry wise multiplication.

When the parameter $|A| > 0.5$, then the operator is expressed by the equation:

$$\vec{X}_{new}(t) = 0.5 \times (\vec{X}_\alpha - \vec{A}_1\vec{D}_\alpha + \vec{X}_\beta - \vec{A}_2\vec{D}_\beta) + rand(size(D)) \oplus Levi(\beta) \quad (7.30)$$

where D is the dimension.

Heidari and Pahlavani (2017), proposed, the Mantegna method Mantegna (1994) to be used to produce Lévy flight during the algorithm process. Consequently, the previous equation can be expressed as:

$$rand(size(Dim)) \oplus Levi(\beta) \sim 0.01\frac{u}{v^{-\beta}}(\vec{X}(t) - \overline{X^\alpha}(t)) \quad (7.31)$$

where u and v are attained from normal distributions as follows:

$$u \sim N(0,\sigma_u^2), v \sim N(0,\sigma_v^2) \quad (7.32)$$

with,

$$\sigma_u = \left[\frac{\Gamma(1+\beta)\sin(\frac{\pi\beta}{2})}{\Gamma(\frac{1+\beta}{2})\beta \times 2^{\frac{\beta-1}{2}}} \right]^{\frac{1}{\beta}}, \sigma_V = 1 \quad (7.33)$$

where,

- Γ represents the conventional gamma function.
- β parameter is a random value inside the [0, 2] interval

The β variable can improve both the exploitation and exploration processes. The Lévy-flight search can prevent the GWO from stagnation.

The greedy selection is adopted with p probability. The operator is expressed by the equation:

$$\vec{X}(t+1) = \begin{cases} \vec{X}(t) & f(\vec{X}_{new}(t)) > f(\vec{X}(t)) \text{ and } r_{new} < p \\ \vec{X}_{new}(t) & \text{otherwise} \end{cases} \quad (7.34)$$

where,

- r_{new} and p represent random values in the interval (0, 1),
- $f(X(t))$ represents the final position fitness value.
- $X_{new}(t)$ represents the new position.
- The p value is defined randomly in the interval [0, 1].

Whale optimization algorithm

Mirjalili and Lewis (2016), proposed a new nature-inspired algorithm called whale optimization algorithm. There are seven different whale species: killer, humpback, Minke, Sei, right, finback, and blue. The authors focused on the humpback whales' hunting strategy. This hunting behavior is named bubble-net feeding method.

Mirjalili and Lewis (2016), mathematically modeled the encircling prey and the spiral bubble-net movement, and the search for prey in order to develop the whale optimization algorithm.

Encircling prey

Humpback whales are able to recognize the prey location and then encircle it. The best whale search agent position (solution) is determined at the beginning as the target prey, then the other whale search agents are trying to update their positions (solutions) with a direction towards the best whale. The authors proposed the following equations to model this behavior:

$$\vec{D} = \left| \vec{C}.\overrightarrow{X^*}(t) - \vec{X}(t) \right| \quad (7.35)$$

$$\vec{X}(t+1) = \overrightarrow{X^*}(t) - \vec{A}\cdot\vec{D} \quad (7.36)$$

where,

- t represents the current iteration.
- $\vec{A}$ and $\vec{C}$ represent the coefficient vectors.

- X^* represents the best solution position vector so far.
- $\vec{X}$ represents the position vector.
- $|\,|$ represents the absolute value.
- $\cdot$ represents an element-wise multiplication.

The $\vec{A}$ and $\vec{C}$ vectors are estimated by using the following equations:

$$\vec{A} = 2\vec{a}\cdot\vec{r} - \vec{a} \tag{7.37}$$

$$\vec{C} = 2\cdot\vec{r} \tag{7.38}$$

where,

- $\vec{a}$ represents a parameter linearly decreased from the value 2 to the value 0 during the iterations.
- $\vec{r}$ represents a random vector in the space [0,1].

The humpback whales also use a Bubble-net attacking strategy. This method can be modelled as follows:

Bubble-net attacking strategy (exploitation phase)

The authors modeled the humpback whales' bubble-net attacking method strategy by proposing two methods:

Shrinking encircling method

This behavior is achieved by decreasing the value of $\vec{a}$ in the equation (2.3). The parameter $\vec{A}$ (fluctuation range) is also decreased by $\vec{a}$ according to the above equation. $\vec{A}$ expresses a random vector which takes values in the space $[-a,a]$.

Spiral updating position

The spiral updating position is performed by calculating the distance between the whale agent located at point (X, Y) and prey located at point (X^*, Y^*). The authors proposed a spiral equation to model the helix-shaped humpback whale movement as follows:

$$\vec{X}(t+1) = \overrightarrow{D'}\cdot e^{bl}\cdot\cos(2\pi l) + \overrightarrow{X^*}(t) \tag{7.39}$$

where,

- $\overrightarrow{D'} = \left|\overrightarrow{X^*}(t) - \vec{X}(t)\right|$ represents the distance of the ith whale from the prey
- $\vec{X}*$ represents the best solution position so far.

- $\vec{X}$ represents the position vector.
- b represents a constant related to the logarithmic spiral shape
- l represents a random number in the interval $[-1,1]$
- $\cdot$ represents an element-wise multiplication

According to the natural behavior of humpback whales, they swim within a shrinking circle around the prey and also in a spiral shape at the same time. The authors, proposed a probability of 50 percent to choose between either method in order to update the whale position by using the equation:

$$\vec{X}(t+1) = \begin{cases} \overrightarrow{X^*}(t) - \vec{A} \cdot \vec{D} & \text{if } p < 0.5 \\ \vec{D'} \cdot e^{bl} \cdot \cos(2\pi l) + \overrightarrow{X^*}(t) & \text{if } p \geq 0.5 \end{cases} \tag{7.40}$$

where p represents a random number in the interval $[0,1]$.

Search for prey (exploration stage)

Furthermore, the humpback whales search for prey randomly. A similar method based on the variation of the $\vec{A}$ vector can be applied in the exploration phase in order to search for prey.

In this phase the position of a whale search agent is updated according to a randomly selected whale search agent, in contradiction to the exploration phase where the best search agent so far matters in the process. $\vec{A} > 1$ is utilized in order to allow the whale optimization algorithm to execute a global search. The proposed equations of the model are as follows:

$$\vec{D} = \left| \vec{C} \cdot \overrightarrow{X_{rand}} - \vec{X} \right| \tag{7.41}$$

$$\vec{X}(t+1) = \overrightarrow{X_{rand}} - \vec{A} \cdot \vec{D} \tag{7.42}$$

where $\overrightarrow{X_{rand}}$ represents a random whale position (solution) vector selected from the current population.

According to the proposed WOA algorithm (Mirjalili and Lewis, 2016), the population whales are initialized with random positions (solutions), the best whale solution, the algorithm parameters and the maximum number of iterations. The whale search agents update their positions according to the two proposed mechanisms: a randomly chosen whale search agent or the best whale position (solution) so far. Parameter "a" is decreased from the value 2 to the value 0 so as to enhance the exploration and the exploitation, accordingly. When the parameter $\vec{A} > 1$ a the random whale search agent is selected. When $\vec{A} < 1$ the best position (solution) is selected for the position update. The value of the p parameter is used to permit WOA algorithm to switch between the two mechanisms: the spiral or the circular movement. Finally, the algorithm terminates when it reaches the termination criterion.

The pseudo code of the WOA algorithm is presented below:

Whale optimization algorithm

```
Initialize the population of whales X, (i = 1, 2, ... , n)

Calculate the fitness of each whale search agent

X* = the best whale search agent

while (1 < MaximumNumberOfiterations)

    for each search agent

        Update A, C, l and p parameters

        If (p<0.5) then

            If |A| < 1

                Update the current search agent position by the eq.(7.35)

            else if (|A| > 1)

                select a random whale whale agent X_rand

                Update the current whale agent position by the eq.(7.42)

            end if

        else if (p>0.5) then

            Update the current whale agent position by the eq.(7.39)

        end if

    end for

    Check if a whale search agent goes away from search space and amend it
```

```
Calculate the fitness of each whale search agent

Update X* if there is a better position (solution)

t=t+1
```

end while

```
Return X*
```

Whale optimization variants

Whale optimization with Lévy flight

Ling et al. (2017), proposed a Lévy flight whale optimization algorithm in order to improve the performance of the algorithm in multi-modal and high-dimensional optimization problems. The authors proposed the incorporation of Lévy flight in the whale optimization algorithm to improve the whale optimization algorithm exploration and the exploitation processes. Also, Lévy flight can improve the diversification of the whale search agents, so as to explore the search space more efficiently. The authors implement Lévy flight to update the humpback whale positions (solutions) according to the following equation:

$$\vec{X}(t+1) = \vec{X}(t) + \mu \, sign[rand - 1/2] \oplus Levy \qquad (7.43)$$

where,

- $\vec{X}(t)$ represents the position (solution) vector $\vec{X}$ at iteration t.
- μ represents a uniformly distributed random number.
- $\oplus$ represents the entry-wise multiplication.
- rand represents a random number in the space [0, 1].
- $sign[rand - 1/2]$ takes only three values: 1, 0, −1.

The Lévy flight provides a random walk and the random step length is expressed by the Lévy distribution which is heavy-tailed:

$$Levy \sim u = t^{-\lambda}, 1 < \lambda \leq 3 \qquad (7.44)$$

The authors used the Mantegna algorithm (Mantegna, 1994) to simulate a λ distribution by producing random steps with length s with the same behavior as the Lévy-flights, according to the equation:

$$s = \frac{\mu}{|v|^{1/\beta}} \qquad (7.45)$$

where,

- s is the step length of the Lévy flight, which is $Levy(\lambda)$
- $\lambda = 1 + \beta$ where $\beta = 1.5$, $\mu = N(0, \sigma_\mu^2)$ and $v = N(0, \sigma_v^2)$ represent normal stochastic distributions with

$$\sigma_\mu = \left[\frac{\Gamma(1+\beta) \times \sin(\pi \times \beta/2)}{\Gamma\big(\big((1+\beta/2)\big)\big) \times \beta \times 2^{(\beta-1)/2}} \right]^{1/\beta} \quad \text{and} \quad \sigma_v = 1 \qquad (7.46)$$

Lévy flight can improve the search capability of the whale optimization algorithm and prevent it from being trapped in a local minimum. The proposed algorithm, improved the experimental results in unimodal and multimodal benchmark functions that were tested.

The pseudocode of the Lévy flight whale optimization algorithm is presented below (Ling et al., 2017):

Lévy Flight Whale optimization algorithm

```
Initialize the population of whales X, (i = 1, 2, ... , n)
Calculate the fitness of each whale search agent
X* = the best whale search agent
while (1 < MaximumNumberOfiterations)
   for each search agent
      Update A, C, l and p parameters
      If (p<0.5) then
         If |A| < 1
            Update the current search agent position by the eq.(7.35)
```

```
    else if (|A| > 1)

        select a random whale whale agent X_rand

        Update the current whale agent position by the eq.(7.42)

        end if

    else if (p>0.5) then

        Update current whale position using Lévy flight by the eq.(7.43)

    end if

    end for

    Check if a whale search agent goes away from search space and amend it

    Calculate the fitness of each whale search agent

    Update X* if there is a better position (solution)

    t=t+1

end while

Return X*
```

Binary whale optimization algorithm

Hussien et al. (2020), proposed a binary version of the whale optimization algorithm by using zero or one to update the whale agent position. The authors proposed changing the whale agent position (solution) according to the distance probability and utilized a transfer function in order to map the distance values to probability values for updating the positions. The sigmoid (S-shaped) and hyperbolic tan (V-shaped) functions to transform the original equations:

$$S(x_i^k(t)) \quad = \frac{1}{1+e^{-d_i^k}(t)} \tag{7.47}$$

$$V(x_i^k(t)) \quad = |\tanh(x_i^k(t))| \tag{7.48}$$

where $d_i^*(t)$ is the particle distance.

$$x_i^{k+1}(t+1) = \begin{cases} 0 & \text{if } rand < S\left(x_i^k(t)\right) \\ 1 & \text{otherwise} \end{cases} \tag{7.49}$$

$$x_i^{k+1}(t+1) = \begin{cases} \left(x_i^k(t)\right)^{-1} & \text{if } rand < V\left(x_i^k(t)\right) \\ x_i^k(t) & \text{otherwise} \end{cases} \tag{7.50}$$

where,

- $x_i^k(t)$ represents the position of the ith particle at iteration t.
- $(x_i^k(t))^{-1}$ represents the complement of $x_i^k(t)$.

Hussien et al. (2020), tested the performance of the proposed binary version of the whale optimization algorithm by using several benchmark functions and engineering design problems. The experimental results showed that the proposed algorithm has a better performance than other optimization algorithms.

Grasshopper optimization algorithm

Saremi et al. (2017), proposed the Grasshopper optimization algorithm, another nature-inspired algorithm. Grasshoppers are insects. The main feature of the Grasshopper swarm in the larval stage is their small steps which make them move slowly. Also, the Grasshopper swarm, moves abruptly in a long-range when the swarm is in its adulthood. There is attraction or repulsion among the Grasshoppers. When a grasshopper is at a specific distance from another grasshopper, there is no attraction or repulsion. This zone is named comfort zone.

Generally, the proposed nature-inspired algorithms use two search strategies: exploration and exploitation. These two operations are performed by grasshoppers in nature.

The authors proposed the following mathematical model to simulate the swarm behavior of the grasshoppers (Topaz et al., 2008):

$$X_i = S_i + G_i + A_i \tag{7.51}$$

where,

- X_i represents the position (solution) of the i_{th} grasshopper.
- S_i represents the social interaction.

- G_i represents the gravitational force exercised on the i_{th} grasshopper.
- A_i represents the wind advection.

By incorporating random behavior to the previous equation, it can be expressed as:

$$X_i = r_1 S_i + r_2 G_i + r_3 A_i \qquad (7.52)$$

where, r_1, r_2, and r_3 represent random numbers in the range [0,1].

$$S_i = \sum_{\substack{j=1 \\ j \neq i}}^{N} s(d_{ij}) \hat{d}_{ij} \qquad (7.53)$$

where,

- d_{ij} represents the distance between i_{th} and j_{th} grasshopper: $d_{ij} = |x_j - x_i|$
- s represents a function for the strength of social forces.

- $\hat{d}_{ij} = \dfrac{x_j - x_i}{d_{ij}}$ represents the vector from i_{th} to j_{th} grasshopper.

The $s(r)$ function, expresses the social forces:

$$s(r) = f e^{\frac{-r}{l}} - e^{-r} \qquad (7.54)$$

where,

- f represents the attraction intensity
- l represents the attraction length scale.

The G parameter in equation (7.51) is estimated by the formula:

$$G_i = -g \hat{e}_g \qquad (7.55)$$

where,

- g represents the gravitational constant.
- $\hat{e}_g$ represents a unitary vector with the direction to the center of earth.

The A parameter in equation (7.51) is estimated by the formula:

$$A_i = u \hat{e}_w \qquad (7.56)$$

where,

- u represents a constant drift.
- $\hat{e}_w$ represents a unitary vector in the wind direction.

By combining the previous equations and substituting the s, G, A parameters in equation (7.51), it can be expressed as follows:

$$X_i = \sum_{\substack{j=1 \\ j\neq i}}^{N} s\left(\left|x_j - x_i\right|\right)\frac{x_j - x_i}{d_{ij}} - g\hat{e}_g + u\hat{e}_w \tag{7.57}$$

where,

- $s(r) = fe^{\frac{-r}{l}} - e^{-r}$
- N represents the number of grasshoppers.

The authors proposed a modified version of the above equation, because in the previous model the grasshopper swarm cannot converge, as follows:

$$X_i^d = c\left(\sum_{\substack{j=1 \\ j\neq i}}^{N} c\frac{ub_d - lb_d}{2} s\left(\left|x_j^d - x_i^d\right|\right)\frac{x_j - x_i}{d_{ij}}\right) + \hat{T}_d \tag{7.58}$$

where,

- ub_d represents the upper bound in the dth dimension,
- lb_d represents the lower bound in the dth dimension
- $s(r) = fe^{\frac{-r}{l}} - e^{-r}$
- $\hat{T}_d$ represents the best solution found so far in the dth dimension.
- c represents a decreasing coefficient parameter to reduce the comfort repulsion and attraction zones.

The first term in the equation simulates the interaction between the grasshoppers and the second term, simulates the inclination of the grasshoppers towards the food source. The adaptive parameter c is similar to the inertial weight (w) in the particle swarm optimization algorithm. This adaptive parameter is used to facilitate the balance between exploration and exploitation. The parameter $s(|x_j - x_i|)$ shows whether a grasshopper must be repelled (exploration) or attracted (exploitation) to the target.

In order to balance the exploration and the exploitation, the parameter c will be decreased while the number of iterations increase to promote exploitation:

$$c = c_{max} - l\frac{c_{max} - c_{min}}{L} \qquad (7.59)$$

where,

- c_{max} represents the maximum value
- c_{min} represents the minimum value
- l represents the current iteration,
- L represents the maximum number of iterations.

Consequently, repulsion makes the grasshopper agents explore the search space, and attraction makes the grasshopper agents exploit regions of the search space. The balance between these two operations: exploration and exploitation, is performed by the adaptive coefficient. The pseudo code of the grasshopper optimization algorithm is as follows (Saremi et al., 2017):

Grasshopper Optimization Algorithm

```
Initialize the grasshopper swarm X (i = 1, 2, ... ,n)

Initialize parameters Cmax, Cmin, and set maximum number of iterations

Estimate the fitness value of every search agent

T=the best grasshopper search agent

while (l < MaximumNumberOfiterations)

   Update c using Eq. (7.59)

   for each grasshopper search agent

      Normalize the distances between the grasshoppers in [1,4}

      Update current grasshopper agent position by the equation (7.58)

      Bring the current grasshopper agent back if it is out of the limits

   end for

   Update T if a better solution exists

   l=l+l

end while

Return T
```

Grasshopper optimization variants

Chaotic grasshopper optimization algorithm

Arora and Anand (2019), proposed a Chaotic grasshopper optimization algorithm. The authors proposed the use of c_1, c_2 instead of the c coefficient, according to the following equation:

$$X_i^d = c_1 (\sum_{j=1, j \neq i}^{N} c_2 \frac{ub_d - lb_d}{2} s(|x_j^d - x_i^d|) \frac{x_j - x_i}{d_{ij}}) + \hat{T}_d \qquad (7.60)$$

The coefficient c_1 is used to balance the exploration and exploitation, and coefficient c_2 is utilized to decrease the attraction, comfort and repulsion zones between the grasshoppers.

Chaotic maps have been used in several optimization algorithms to improve their performance. Arora and Anand (2019), proposed adapting the c_1 and c_2 coefficients by using several chaotic maps.

Adapting c_1 coefficient with chaotic maps

The coefficient c_1 is used to balance the exploration and exploitation. Arora and Anand (2019), proposed several chaotic maps to adapt the c_1 coefficient in the interval [0, 1] according to the equation:

$$X_i^d = c_1(t) (\sum_{j=1, j \neq i}^{N} c_2 \frac{ub_d - lb_d}{2} s(|x_j^d - x_i^d|) \frac{x_j - x_i}{d_{ij}}) + \hat{T}_d \qquad (7.61)$$

where,

- $c_1(t)$ represents the chaotic map value in the t_{th} iteration
- X_i represents the i_{th} grasshopper.

Adapting c_2 coefficient with chaotic maps

The coefficient c_2 is utilized to decrease the attraction, comfort and repulsion zones between the grasshoppers. The c_2 coefficient is adapted in the interval [0, 1] by using chaotic maps according to the equation:

$$X_i^d = c_1 (\sum_{j=1, j \neq i}^{N} c_2(t) \frac{ub_d - lb_d}{2} s(|x_j^d - x_i^d|) \frac{x_j - x_i}{d_{ij}}) + \hat{T}_d \qquad (7.62)$$

where,

- $c_2(t)$ represents the chaotic map value in the t_{th} iteration.
- X_i represents the i_{th} grasshopper.

Improved grasshopper optimization algorithm

Luo et al. (2018), proposed an improved grasshopper optimization algorithm, by using three methods: Levy flight, opposition-based learning, and Gaussian mutation. These strategies were used for better balancing the two main operations: exploration and exploitation.

Gaussian mutation

The Gaussian mutation method was proposed by Luo et al. (2018), in order to increase the grasshopper optimization algorithm population diversity in order to increase the search capability of the algorithm for finding the global optimum. The modified equation of the original grasshopper optimization algorithm is:

$$X_i^d = c \left(\sum_{\substack{j=1 \\ j \neq i}}^{N} c \frac{ub_d - lb_d}{2} s\left(\left|x_j^d - x_i^d\right|\right) \frac{x_j - x_i}{d_{ij}} \right) \oplus G(\alpha) + \hat{T}_d \quad (7.63)$$

Levy flight

After the grasshopper position update, Levy flight method is proposed by the authors to produce a new solution, by using the following equations:

$$X_i^{levy} = X_i^* + rand(d) \oplus levy(\beta) \quad (7.64)$$

$$X_i^{t+1} = \begin{cases} X_i^{levy} & fitness\left(X_i^{levy}\right) > fitness\left(X_i^*\right) \\ X_i^* & otherwise \end{cases} \quad (7.65)$$

where,

- X_i^* represents the new position (solution) of the i_{th} grasshopper.
- rand(d) represents a d-dimensional vector randomly generated in [0,1].

Opposition-based learning

The authors, also used the opposition-based learning method in order to produce the oppositional population which corresponds to the current grasshopper population after the positions of the grasshopper are updated. This method helps the algorithm to search more effectively and enhance the algorithm exploration ability. The grasshopper oppositional population can be produced by the following equation:

$$X_i^{op} = LB + UB - T + r\left(T - X_i\right) \tag{7.66}$$

where,

- X_i^{op} represents the position of the i_{th} opposite grasshopper.
- LB and UB represent the lower and the upper bounds correspondingly.
- T represents the position (solution) of the best grasshopper.
- r represents a random vector in the space $(0,1)$.
- X_i represents the position (solution) vector of the i_{th} grasshopper.

The proposed method for improving the original grasshopper optimization algorithm by using these three strategies was tested in several benchmark functions. The experimental results showed an improved accuracy compared to other swarm optimization algorithms such as particle swarm optimization and differential evolution.

References

Altan, A., Karasu, S. and Zio, E. (2021). A new hybrid model for wind speed forecasting combining long short-term memory neural network, decomposition methods and grey wolf optimizer. *Applied Soft Computing*, 100: 106996.

Arora, S. and Anand, P. (2019). Chaotic grasshopper optimization algorithm for global optimization. *Neural Computing and Applications,* 31(8): 4385–4405. doi:10.1007/s00521-018-3343-2.

Emary, E., Zawbaa, H. M., Grosan, C. and Hassenian, A. E. (2015). Feature subset selection approach by gray-wolf optimization. pp. 1–13. *In: Afro-European Conference for Industrial Advancement.* Springer, Cham.

Emary, E., Zawbaa, H. M. and Hassanien, A. E. (2016). Binary grey wolf optimization approaches for feature selection. *Neurocomputing*, 172: 371–381.

Ghalambaz, M., Yengejeh, R. J. and Davami, A. H. (2021). Building energy optimization using Grey Wolf Optimizer (GWO). *Case Studies in Thermal Engineering*, 27: 101250.

Heidari, A. A. and Pahlavani, P. (2017). An efficient modified grey wolf optimizer with Lévy flight for optimization tasks. *Applied Soft Computing,* 60: 115–134. doi:https://doi.org/10.1016/j.asoc.2017.06.044.

Helmi, A. M., Al-Qaness, M. A., Dahou, A., Damaševičius, R., Krilavičius, T. et al. (2021). A novel hybrid gradient-based optimizer and grey wolf optimizer feature selection method for human activity recognition using smartphone sensors. *Entropy*, 23(8): 1065.

Hu, P., Pan, J. S. and Chu, S. C. (2020). Improved binary grey wolf optimizer and its application for feature selection. *Knowledge-Based Systems*, 195: 105746.

Hussien, A. G., Hassanien, A. E., Houssein, E. H., Amin, M., Azar, A. T. et al. (2020). New binary whale optimization algorithm for discrete optimization problems. *Engineering Optimization,* 52(6): 945–959. doi:10.1080/0305215X.2019.1624740.

Jayabarathi, T., Raghunathan, T., Adarsh, B. R. and Suganthan, P. N. (2016). Economic dispatch using hybrid grey wolf optimizer. *Energy,* 111: 630–641.

Li, I., Fu, Y., Fung, T. C., Qu, H., Lau, A. K. et al. (2021). Development of a back-propagation neural network and adaptive grey wolf optimizer algorithm for thermal comfort and energy consumption prediction and optimization. *Energy and Buildings,* 253: 111439.

Ling, Y., Zhou, Y. and Luo, Q. (2017). Lévy flight trajectory-based whale optimization algorithm for global optimization. *IEEE Access,* 5: 6168–6186. doi:10.1109/ACCESS.2017.2695498.

Luo, J., Chen, H., Zhang, Q., Xu, Y., Huang, H. et al. (2018). An improved grasshopper optimization algorithm with application to financial stress prediction. *Applied Mathematical Modelling,* 64: 654–668. doi:https://doi.org/10.1016/j.apm.2018.07.044.

Mantegna, R. N. (1994). Fast, accurate algorithm for numerical simulation of Levy stable stochastic processes. *Physical Review E,* 49(5): 4677.

Mirjalili, S., Mirjalili, S. M. and Lewis, A. (2014). Grey Wolf Optimizer. *Advances in Engineering Software,* 69: 46–61. doi:https://doi.org/10.1016/j.advengsoft.2013.12.007.

Mirjalili, S. and Lewis, A. (2016). The whale optimization algorithm. *Advances in Engineering Software,* 95: 51–67.

Nadimi-Shahraki, M. H., Taghian, S. and Mirjalili, S. (2021). An improved grey wolf optimizer for solving engineering problems. *Expert Systems with Applications,* 166: 113917.

Saremi, S., Mirjalili, S. and Lewis, A. (2017). Grasshopper optimisation algorithm: theory and application. *Advances in Engineering Software,* 105: 30–47.

Thakur, N., Awasthi, Y. K. and Siddiqui, A. S. (2021). Reliability analysis and power quality improvement model using enthalpy based grey wolf optimizer. *Energy Systems,* 12(1): 31–59.

Topaz, C. M., Bernoff, A. J., Logan, S. and Toolson, W. (2008). A model for rolling swarms of locusts. *The European Physical Journal Special Topics,* 157(1): 93–109.

Wong, L. I., Sulaiman, M. H., Mohamed, M. R. and Hong, M. S. (2014, December). Grey Wolf Optimizer for solving economic dispatch problems. pp. 150–154. *In: 2014 IEEE International Conference on Power and Energy (PECon).* IEEE.

Zhang, Z. and Hong, W.-C. (2021). Application of variational mode decomposition and chaotic grey wolf optimizer with support vector regression for forecasting electric loads. *Knowledge-Based Systems,* 228: 107297. doi:https://doi.org/10.1016/j.knosys.2021.107297.

CHAPTER 8

Machine Learning Optimization Applications

This chapter is devoted to machine learning optimization applications, such as artificial neural network optimization. A neural network can be optimized by using swarm and evolutionary optimization methods in several ways, such as: weight optimization, structure optimization, feature selection with evolutionary algorithms. Illustrative real-world examples are presented with real datasets, by using several methods to optimize a neural network such as: genetic algorithm, particle swarm optimization and ant colony optimization.

Artificial neural networks

The Artificial Neural Networks (ANNs) are artificial computational systems that simulate the neural architecture of a human brain. A neural network elaborates the data from the input variables. The input data traverse through the neuron connections and after being processed, the output results are estimated according to the input parameters (Basheer and Hajmeer, 2000; Panchal et al., 2011).

In a feed forward multilayer neural network, the neurons are connected only to a forward direction and the typical architecture consists of an input layer, one or more hidden layers and an output layer. Each layer is developed by a number of neurons (Castro et al., 2000; Honik, 1991; Yam and Chow, 2001).

Several kinds of activation functions can be utilized in a multilayer perceptron. When the sigmoid function is implemented, then the output of the hidden layer neuron *j* can be estimated by using the following equation:

$$S_j = \sum_{i=1}^{n} w_{ij} I_i + \beta_j \tag{8.1}$$

$$f_j(x) = \frac{1}{1 + e^{-S_j}} \tag{8.2}$$

where,

- S_j represents the weighted summation of the input hidden layer neuron *j*
- I_i represents the input variable *i*
- w_{ij} represents the connection weights between I_i and the hidden neuron *j*.

After estimating the output of every neuron in the hidden layer, the output of the neural network can be found by the following equation:

$$\hat{y}_k = \sum_{i=1}^{m} W_{kj} f_i + \beta_k \tag{8.3}$$

Weight optimization of a neural network

The weight optimization can be implemented by using swarm and evolutionary optimization methods (Ilonen et al., 2003; Socha and Blum, 2007). A fitness function is set in order to be used to train the neural network. The most common objective function that is used is an error metric. In regression problems the most common metrics are: the mean squared error (MSE) or root mean squared error (RMSE) or mean absolute error (MAE) which can be utilized as objective functions.

In classification problems other metrics can be utilized, such as: accuracy, precision, recall, F1 and kappa. After the selection of the objective function, the swarm topology is developed. Each particle position represents a solution to the objective function optimization problem and the neural network weight vector is optimized until the swarm finds the best solution for the objective function.

Topology optimization of a neural network

Swarm intelligence strategies can be also implemented to optimize several parameters of a neural network such as the number of the hidden neurons in the hidden layer. Each particle represents the number of hidden neurons in the hidden layer.

The objective function should be set for the swarm intelligence method or for the genetic algorithm in order to optimize the neural network. The most common objective function that is utilized is an error metric. In regression problems and in classification problems several error metrics can be used as objective functions (Benardos and Vosniakos, 2007; Mirjalili et al., 2012; Zhang and Shao, 2000).

According to several researches, swarm intelligence techniques have a better performance in artificial neural network optimization than other optimization algorithms, such as genetic algorithm (Anand and Suganthi, 2020; Moayedi et al., 2019), or artificial bee colony (Koopialipoor et al., 2019), and non-dominated sorting genetic algorithm II (NSGA-II) (Anand and Suganthi, 2020; Nourbakhsh et al., 2011).

Neural network training with PSO, ACO, GA

In this experiment, the iris dataset is utilized in order to train an artificial neural network (ANN) with 9 hidden neurons in the hidden layer. The Iris dataset was used for the experiments. The neural network has four inputs. The Iris dataset has five attributes, sepal length, sepal width, petal width and petal length, and an attribute which is corresponding to the class of the flower: Iris setosa, Iris virginica and Iris versicolor.

Experimental setup

Genetic algorithm parameters

Maximum Number of generations (Iterations) MaxIt: 1000

Chromosome Population: 200

Number of genes in each chromosome: 75

Selection: Roulette wheel

Crossover type: single point

Crossover probability	1
Mutation:	Uniform
Mutation probability:	0.01

PSO parameters

Maximum Number of Iterations MaxIt:	1000
Swarm topology:	Fully connected
Swarm Population:	200
Inertia weight:	0.33
coefficient parameter of personal learning:	$c_1 = 1.0$
coefficient parameter of global learning	$c_2 = 1.5$

ACO parameters

Maximum Number of generations (Iterations) MaxIt:	1000
Ant Population:	200
Initial pheromone value:	Initial_pher = 1e-6
pheromone update quantity constant:	$Q = 20$
Exploration constant:	$q = 1$
Global pheromone decay rate:	$\alpha = 0.9$
Local pheromone decay rate:	$r = 0.5$
Evaporation Rate:	$\rho = 0.05$

Experimental results

According to the results the best models were the PSO-ANN and the GA-ANN models compared to the ACO-ANN model. The PSO-ANN model has much faster convergence compared to the other models.

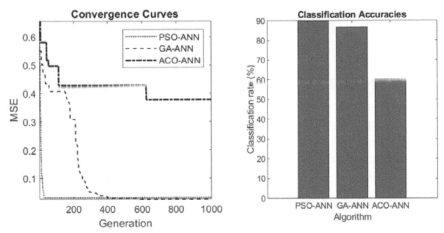

Figure 8.1. Convergence curves and classification accuracies of the models.

Feature selection with swarm intelligence and genetic algorithm

Problem definition

The application of machine learning techniques has been studied by several researchers in the field of energy forecasting (Azadeh et al., 2013; Candanedo et al., 2017; Ekonomou, 2010; Kouziokas, 2019). Candanedo et al. (2017) proposed data driven prediction models regarding the energy use of appliances in low-energy houses by implementing machine learning models such as support vector machines, random forest and gradient boosting machines. Moldovan and Slowik (2021), developed forecasting models for the energy consumptions of appliances by combining multi-objective binary grey wolf optimization method with machine learning methods such as: support vector machines, random forest, extra trees and K-nearest neighbours.

In this section, the appliances energy prediction in smart houses with IoT devices is performed, by using artificial neural network models combined with particle swarm optimization and genetic algorithm.

Data analysis in machine learning

The data analysis methodology in machine learning mainly includes several stages such as: data collection, data cleansing, normalization and preparation, feature selection, development of the machine learning models by testing several architectures and finally model evaluation by using the appropriate evaluation metrics.

Energy consumption dataset

The appliances' energy prediction dataset was acquired by the machine learning repository[1] (Candanedo et al., 2017). The dataset consists of 19735 rows with several characteristics. The dataset was developed by the authors by measuring the values every 10 minutes for 4.5 months in a low energy house with IoT devices. The description of the dataset is presented in the next table.

Data pre-processing

The data was pre-processed so as to erase incoherencies and null values. The output parameter for the machine learning methods is the "appliances' energy consumption". The input vector is all the other variables of the dataset. The dataset was normalized by using the min-max normalization method.

Normalization

The min-max normalization method was used to transform the values of the dataset in the interval [0,1], according to the following equation:

$$n_i = \frac{x_i - \min(x)}{\max(x) - \min(x)} \qquad (8.4)$$

where,

- n_i illustrates the ith normalized value.
- x_i illustrates the ith sample value.
- $\min(x)$ illustrates the minimum value in the dataset.
- *max*(x) illustrates the maximum value in the dataset.

---

[1] https://archive.ics.uci.edu

Table 8.1. Data variables and description.

Data variables	Units	Number of Variable
Appliances' energy consumption	Wh	1
Light energy consumption	Wh	2
T1, Temperature in kitchen area	°C	3
RH1, Humidity in kitchen area	%	4
T2, Temperature in living room area	°C	5
RH2, Humidity in living room area	%	6
T3, Temperature in laundry room area	°C	7
RH3, Humidity in laundry room area	%	8
T4, Temperature in office room	°C	9
RH4, Humidity in office room	%	10
T5, Temperature in bathroom	°C	11
RH5, Humidity in bathroom	%	12
T6, Temperature outside the building (north side)	°C	13
RH6, Humidity outside the building	%	14
T7, Temperature in ironing room	°C	15
RH_7, Humidity in ironing room	%	16
T8, Temperature in teenager room 2	°C	17
RH_8, Humidity in teenager room 2	%	18
T9, Temperature in parents room	°C	19
RH_9, Humidity in parents room	%	20
To, Temperature outside (from Chievres weather station)	°C	21
Pressure (from Chievres weather station)	mm Hg	22
RH_out, Humidity outside (from Chievres weather station)	%	23
Wind speed (from Chievres weather station)	m/s	24
Visibility (from Chievres weather station)	km	25
Tdewpoint (from Chievres weather station)	°C	26
rv1, Random variable 1	nondimensional	27
rv2, Random variable 2	nondimensional	28

Processing dataset outliers

The selected dataset appears extremely high and low values which can affect the generalization ability of the machine learning models. The best method for developing improved machine learning models is to remove the extreme values. In order to discover the dataset outliers regarding the output, some statistic measures will be used such as, the sample mean, the standard deviation, the kurtosis and the skewness. The sample mean $\overline{X}$ is calculated, according to the following equation:

$$\overline{X} = \frac{\sum_{i=1}^{n} X_i}{n} \tag{8.5}$$

The sample standard deviation is calculated, according to the following equation:

$$S = \left[\sum_{i=1}^{n} \left(X_i - \overline{X} \right)^2 / (n-1) \right]^{1/2} \tag{8.6}$$

The skewness is a statistical measure of symmetry of the probability distribution. The larger the absolute value of the skewness is, the greater deviation has from the normal distribution. The skewness is calculated by the following equation (Wang et al., 2018):

$$Skewness = \frac{\sum_{i=1}^{N} (X_i - \overline{X})^3}{(n-1)S^3} \tag{8.7}$$

The Kurtosis is a statistical measure which shows if the data are heavy-tailed or light-tailed compared to a normal distribution and it is calculated, according to the following equation:

$$Kurtosis = \frac{\sum_{i=1}^{N} (X_i - \overline{X})^4}{(n-1)S^4} \tag{8.8}$$

The kurtosis for the standard normal distribution is equal to 3. Some sources in the literature propose the following equation for calculating kurtosis, referred also as "excess kurtosis" (Wang et al., 2018):

$$Kurtosis = \frac{\sum_{i=1}^{N} (X_i - \overline{X})^4}{(n-1)S^4} - 3. \tag{8.9}$$

If the distribution is normal then the rule 1.5*IQR can be used, where IQR represent the interquartile range between Q_1 and Q_2 quartiles of the distribution. The lower and the upper bounds can be calculated by using the formulas:

$$lower \ bound \ = \ Q_1 - 1.5 * IQR \qquad (8.10)$$

$$upper \ bound \ = \ Q_2 + 1.5 * IQR \qquad (8.11)$$

In this case, the kurtosis of the sample is –0.0023 and the skewness is positive 0.7903 and differs from the normal distribution. So, the $2*S$ rule will be used to remove the outliers. According to this rule, the sample standard deviation S of the target values is estimated and the data outlier instances are removed by calculating the lower and the upper bounds according to the formulas:

$$lower \ bound \ = \ \overline{X} - 2 * S \qquad (8.12)$$

$$upper \ bound \ = \ \overline{X} + 2 * S \qquad (8.13)$$

where,

- $\overline{X}$ represents the sample mean.
- S illustrates the sample standard deviation.

The sample mean was 24.75 and the sample standard deviation was 64.68, so according to the above equations the lower bound was 15.18 and the upper bound 114.18.

Cost-based feature selection with swarm intelligence

The cost function is generally utilized to evaluate the performance of a Machine Learning model. There are several cost functions such as, Mean Absolute Error (MAE), Mean Squared Error (MSE), Root Mean Squared Error (RMSE), Root Mean Squared Logarithmic Error (RMSLE). The cost-based feature subset selection by using swarm and evolutionary intelligence is based on the combination of a classifier or regressor which is evaluated for its performance by a cost function, like MSE, RMSE in regression or accuracy, F1 score, etc. in classification and a swarm or evolutionary optimization method which operates as an optimization stochastic method exploring the search space in order to find the optimal subset of features which yield the best results. The mean squared error (MSE) is expressed by the equation:

$$MSE = \frac{\sum_{n}^{i=1}(Y_i - \hat{Y}_i)^2}{n} \tag{8.14}$$

where:

- Y_i represents the actual value.
- $\hat{Y}_i$ represents the predicted value.
- n represents the number of measurements (instances).

Correlation-based feature selection with swarm intelligence

Since, cost-based methods are much more expensive and highly time-consuming methods to be used in combination with swarm intelligence subset search methods, usually a correlation-based feature subset selection method is utilized. Hall (1999), proposed *Merit*, as a measure of correlation. The Merit is similar to Pearson's correlation equation, and can be used to evaluate the correlation-based feature selection (CFS) algorithm. In order to define the subsets containing the highly class-correlated features and not uncorrelated to each other, the merit measure was proposed by (Hall, 1999):

$$Merit_s = \frac{k\overline{r_{cf}}}{\sqrt{k + k(k-1)\overline{r_{ff}}}} \tag{8.15}$$

where:

- $Merit_s$ expresses the merit of a subset S which contains k features
- $\overline{r_{cf}}$ indicates the average correlation with the class
- $\overline{r_{ff}}$ indicates the average intercorrelation.

The subset which has the highest value of merit is selected. Swarm intelligence algorithms are applied as subset search methods in combination with CFS to improve the efficiency of the CFS algorithm.

Experimental setup

In this experiment, the CFS merit was selected as the fitness function for the swarm and evolutionary algorithms. Several swarm and evolutionary-based feature selection methods were utilized: genetic algorithm, geometric particle swarm optimization, evolutionary algorithm, chaotic harmony search and chaotic Cuckoo search.

Genetic algorithm

GA algorithm parameters

Population size:	20
Number of generations.	20
Selection:	Roulette wheel
Crossover type:	single point
Mutation:	Uniform
Probability of crossover:	0.6
Probability of mutation:	0.033

Genetic Algorithm results

The fitness values of the chromosomes in the last generation of the genetic algorithm are illustrated in the next table.

Table 8.2. Genetic algorithm results of the 20th generation.

Chromosome	Fitness value	Subset
1	**0.43273**	**1 4 16 17**
2	0.43273	1 4 16 17
3	0.39589	1 2 4 10 16 17 22
4	0.33461	1 16 17 21 24
5	0.39	1 4 16 17 24
6	0.43273	1 4 16 17
7	0.43273	1 4 16 17
8	0.35691	1 4 10 14 16 17 21
9	0.38071	1 4 14 16 17 19
10	0.37789	1 4 15 16 17
11	0.37649	1 2 3 4 16 17
12	0.42288	1 4 16 17 22
13	0.43273	1 4 16 17
14	0.32487	1 4 16 25
15	0.40494	1 4 16 17 21
16	0.43273	1 4 16 17
17	0.39774	1 4 16
18	0.33358	1 4 8 11 15 17
19	0.42155	1 4 17
20	0.37404	1 2 4 16 17 22 25

According to the best fitness value (the biggest) the selected attributes in the feature selection with genetic algorithm were five: 1, 4, 16, 17:

- Lights
- T2
- T8
- RH_8

Geometric PSO

GPSO parameters

Geometric PSO (GPSO) was used for the feature selection as proposed by Moraglio et al. (2007). The parameters for the experiments were the following.

Population size: 20

Swarm topology: Fully connected

Number of iterations: 20

Coefficient w_1: $w_1 = 0.33$

Coefficient w_2: $w_2 = 0.33$

Coefficient w_3: $w_3 = 0.34$

Crossover type: three-parent mask-based

Mutation type: bit-flip

Probability of mutation: 0.01

GPSO results

The fitness values of the GPSO particles in the last generation of the GPSO algorithm are illustrated in the next table.

According to the best fitness value (the biggest) the selected attributes in the feature selection with GPSO algorithm were: 1, 2, 16, 17, 22

- Lights
- T1
- T8

Table 8.3. GPSO algorithm results of the 20th iteration.

Particle	Fitness value	Subset
1	0.41967	1 2 16 17 22
2	0.41967	1 2 16 17 22
3	0.41967	1 2 16 17 22
4	0.41967	1 2 16 17 22
5	0.41967	1 2 16 17 22
6	0.41967	1 2 16 17 22
7	0.41967	1 2 16 17 22
8	0.41967	1 2 16 17 22
9	0.41967	1 2 16 17 22
10	0.41967	1 2 16 17 22
11	0.41967	1 2 16 17 22
12	0.38319	1 2 6 11 16 17 22
13	0.39684	1 2 6 17 22
14	0.41967	1 2 16 17 22
15	0.41967	1 2 16 17 22
16	0.41967	1 2 16 17 22
17	0.41967	1 2 16 17 22
18	0.41967	1 2 16 17 22
19	0.3875	1 2 7 16 17 22
20	0.41967	1 2 16 17 22

- RH_8
- RH_out

Chaotic harmony search

Algorithm parameters

Initial population size of new harmonies: 20

Number of iterations: 20

Mutation: bit-flip

Mutation probability: 0.01

Harmony Memory Size: 25

Harmony Memory Accepting Rate: $r_{accept} = 0.9$

Initial Pitch Adjustment Rate (PAR): $r_{pitch} = 0.4$

Chaotic method for PAR: Logistic map

Logistic map coefficient: 4.0

PAR Minimum Bound (PAR_{min}):   0.4

PAR Maximum Bound (PAR_{max}):   0.9

Pitch Bandwidth: $b = 0.02*(PAR_{max} - PAR_{min})$

Chaotic harmony search results

The fitness values of the harmonies in the last generation of the chaotic harmony search algorithm with mutation algorithm are illustrated in the next table.

Table 8.4. Chaotic harmony search with mutation algorithm results of the 20th iteration which had the optimal results.

Harmony	Fitness value	Subset
1	0.27533	7 8 9 12 17 22 24
2	0.33034	1 3 4 12 17 21
3	0.35268	1 4 8 10 13 17
4	0.32453	1 4 6 8 15 21 22 27
5	0.32449	1 4 12 17 18 21 26
6	0.36828	1 4 8 12 15 16 17 21
7	0.33535	1 4 8 12 17 18 21 24
8	0.33596	4 13 17 21
9	0.33432	4 13 17
10	**0.40103**	**1 4 12 17**
11	0.30418	1 12 18
12	0.28135	1 8 18
13	0.3508	4 12 17
14	0.31567	1 4 12 17 25
15	0.33251	1 3 4 12 17 22 24
16	0.3482	4 12 17 22
17	0.37476	1 4 6 8 17 21 22
18	0.31707	1 4 12 18
19	0.32146	4 8 12 14 17
20	0.34047	4 8 12 17

According to the best fitness value (the highest), the selected attributes in the feature selection with Chaotic Harmony Search with mutation algorithm were: 1, 4, 12, 17:

- Lights
- T2
- T6
- RH_8

Chaotic Cuckoo Search

Algorithm parameters

Initial population size:	20
Number of iterations:	20
Mutation:	bit-off
Mutation probability:	0.01
Chaotic method for Pa:	logistic map
Logistic map coefficient:	4.0
Initial discovery rate Pa:	0.25
Sigma Rate:	0.70

Chaotic Cuckoo Search results

The fitness values of the harmonies in the best generation (20th) of the chaotic Cuckoo Search algorithm with mutation are illustrated in the next table.

According to the best fitness value (the highest), the selected attributes in the feature selection with Chaotic Cuckoo Search with mutation algorithm were: 1,16,22:

- Lights
- T8
- RH_out

Table 8.5. Chaotic Cuckoo search with mutation algorithm results of the 20th iteration which had the optimal results.

Harmony	Fitness value	Subset
1	0.3874	1 10 16 22
2	0.41886	1 16 22
3	0.41886	1 16 22
4	0.41886	1 16 22
5	0.41886	1 16 22
6	0.41886	1 16 22
7	0.41886	1 16 22
8	0.41886	1 16 22
9	0.36199	1 16 18 22
10	0.41886	1 16 22
11	0.3649	1 14 16 22
12	0.38557	1 6 16 22
13	0.3874	1 10 16 22
14	0.41886	1 16 22
15	0.41886	1 16 22
16	0.41886	1 16 22
17	0.38557	1 6 16 22
18	0.41886	1 16 22
19	0.39074	1 5 16 22
20	0.41886	1 16

Evolutionary algorithm

EA parameters

Population size: 20

Number of generations: 20

Mutation: bit-flip

Mutation probability: 0.10

Crossover: Simplex crossover (Tsutsui et al., 1999)

Crossover probability: 0.60

Selection operator: tournament selection

EA results

The fitness values of the chromosomes in the 18th generation of the evolutionary algorithm with mutation algorithm are illustrated in the next table.

Table 8.6. Evolutionary algorithm results of the 18th generation.

Chromosome	Fitness value	Subset
1	0.3860	1 2 4 5 10 12 16 17 19 22
2	0.3771	1 2 4 10 12 16 17 19 22 26
3	0.3127	1 2 3 4 10 11 12 13 14 16 17 21 22 25 26 27
4	0.3327	1 2 4 5 6 10 12 14 16 17 18 19 22 23 24 26
5	0.3513	1 2 4 5 10 12 16 17 19 23 26
6	0.3414	1 4 10 11 12 13 15 16 17 19 26
7	0.3501	1 2 4 12 13 16 17 21 22 25 26
8	0.3425	1 2 3 4 11 12 13 16 17 20 21 22 24 26
9	0.3442	1 2 6 10 11 12 13 16 17 19 22 23 26
10	0.3562	1 4 5 12 16 17 18 21 22 26
11	**0.3929**	**1 2 4 5 12 16 17 19 22**
12	0.3663	1 2 4 5 12 16 17 19 21 26
13	0.3249	1 2 4 5 6 7 11 12 13 17 20 22 25
14	0.3470	2 4 5 12 13 16 17 21 22 26
15	0.3684	1 2 4 10 12 16 17 19 20 22 26
16	0.3191	1 6 12 16 17 19 22 23 25 26
17	0.3367	1 2 4 5 10 11 12 14 16 19 21 22 25
18	0.3616	1 2 4 12 13 15 16 17 21 22 26
19	0.3434	1 2 4 5 10 12 13 16 17 18 22 24 26
20	0.3156	1 2 9 10 12 13 16 17 18 19 21 25

According to the best fitness value, the evolutionary algorithm selected the features: 1, 2, 4, 5, 12, 16, 17, 19, 22:

- Lights
- T1
- T2
- RH_2
- T6
- T8
- RH_8

- RH_9
- RH_out

Predictions with reduced features, SVM and random forest

The Random Forest (Breiman, 1996, 1999, 2001) and the Support Vector Machines were used to predict the output, the energy appliances measured by IoT devices by using the input vector of 27 features and compare the predictability with the models produced by using the reduced features as inputs according to the swarm intelligence and evolutionary models. The 10-fold cross validation as a validation technique in order to discover the best model. The RMSE was used as error metric.

The Support Vector Machines were proposed by Vapnik (1995). Support Vector Machines for Regression or Support Vector Regression (SVR) use an ε-insensitive loss function for solving the regression problems. The generic estimating SVM function in regression is expressed by the following equation:

$$f(x) = (w \cdot \Phi(x)) + b \qquad (8.16)$$

where $w \subset R^n$, $b \subset R$ and the Φ represents a non-linear transformation from R^n to high dimensional space. There are several types of kernel functions in Support Vector Machines, such as, linear, Gaussian, polynomial and RBF kernel.

The Lib-SVM library (Chang and Lin, 2001, 2011) was utilized which implements the SMO (Sequential Minimal Optimization) algorithm for the support vector machines (SVMs). The epsilon-Support Vector Regression was implemented and the hyperparameters were: RBF kernel, cost $C = 100$, gamma $\gamma = 0.01$, and epsilon e = 0.001. In the random forest method, the number of trees were 100 and the maximum depth 0.

The comparison table of the comparative results shows that the best swarm intelligence-based method used for feature selection was the evolutionary algorithm with nine input features which has achieved forecasting results very close to the ones with all 27 input features.

Table 8.7. Model results.

Model	Features	Random Forest *RMSE error using 10-Fold Cross-Validation*	Lib-SVM **C = 100, γ = 0.01, e = 0.001, RBF kernel** *RMSE error using 10-Fold Cross-Validation*
Genetic algorithm	(1, 4, 16, 17)	14.59	19.16
Geometric PSO	(1, 2, 16, 17, 22)	13.63	19.20
Chaotic Harmony Search	(1, 4, 12, 17)	14.89	19.18
Chaotic Cuckoo Search	(1, 16, 22)	19.04	19.34
Evolutionary Algorithm	(1, 2, 4, 5, 12, 16, 17, 19, 22)	12.04	18.84
Lib-SVM	All features	-	17.58
Random Forest	All features	11.47	-

Crime forecasting with PSO-SVM

The forecasting models were developed by using Support Vector Machines optimized by PSO. The crime dataset acquired by the official site of the City of Chicago: (data.cityofchicago.org). The Crime Rate represents the number of crimes committed in a day per 200.000 residents. The unemployment data for the city acquired by the official site of the United States Department, Bureau of Labor Statistics (www.bls.gov).

The 10-fold cross validation was used prevent the models from overfitting and every fold has the same sequence of the data instances of the original dataset since it is a time series problem. Daily crime data was utilized for the city of Chicago from 2012 till 2015. Also, unemployment data was taken as an input factor since unemployment affects the crime rates.

The input vector X_t of the forecasting model in time t includes: the unemployment rate UN_{t-1}, and the Crime Rate CR_{t-1}, of the previous day. The time step is the day.

$$X_t = (CR_{t-1}, UN_{t-1}) \tag{8.17}$$

The target vector T_t of the machine learning model in time t is the one step ahead value of the input vector parameters:

$$T_t = CR_t \tag{8.18}$$

The PSO was used as a hyperparameter optimization method. The Root Mean Square Error (RMSE) was used in order to evaluate the prediction error and to measure the performance of the PSO optimized machine learning model.

The most common PSO hyperparameters were utilized: population size: $P = 50$, inertia weight $w = 0.99$, personal learning coefficient $c_1 = 2.0$, global learning coefficient $c_2 = 2.0$ and the fully connected topology. The PSO algorithm was utilized to investigate the search space in order to find the best hyperparameter values. The RBF SVM kernel was used since, generally, it is considered to produce the optimal results. The hyperparameters that were tuned were: the cost C, the gamma (γ) parameter, the epsilon parameter and the choice if the input vector was standardized or not. Hyperparameters search ranges are presented in the next table.

The optimal SVM hyperparameter architecture found by the PSO algorithm within 30 iterations was: Cost C = 1.1426, gamma γ = 3.5907, epsilon ε = 0.023833, standardize data: false. The RMSE was 5.4882.

Table 8.8. Hyperparameters search range by applying PSO.

Hyperparameter	Lower bound	Upper bound
Cost C	0.001	1000
gamma γ	0.001	1000
epsilon ε	0.010	100
standardize	true	false

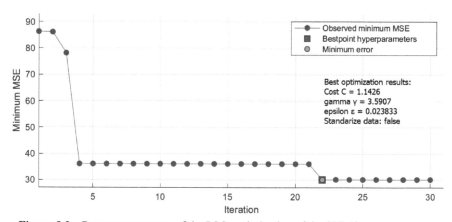

Figure 8.2. Convergence curve of the PSO optimization of the SVM hyperparameters.

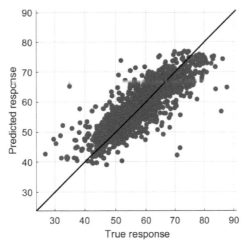

Figure 8.3. Comparison plot of the true and the predicted value of the crime rate (number of crimes per 200.000 residents).

References

Anand, A. and Suganthi, L. (2020). Forecasting of electricity demand by hybrid ANN-PSO models. pp. 865–882 *In: Deep Learning and Neural Networks: Concepts, Methodologies, Tools, and Applications*. IGI Global.

Azadeh, A., Babazadeh, R. and Asadzadeh, S. M. (2013). Optimum estimation and forecasting of renewable energy consumption by artificial neural networks. *Renewable and Sustainable Energy Reviews*, 27: 605–612.

Basheer, I. A. and Hajmeer, M. (2000). Artificial neural networks: fundamentals, computing, design, and application. *Journal of Microbiological Methods*, 43(1): 3–31.

Benardos, P. G. and Vosniakos, G. C. (2007). Optimizing feedforward artificial neural network architecture. *Engineering Applications of Artificial Intelligence*, 20(3): 365–382.

Breiman, L. (1996). Bagging predictors. *Machine learning*, 24(2): 123–140.

Breiman, L. (1999). *Random Forests-Random Features (# 567)*. Technical report, Dept. of Statistics, Univ. of California, Berkeley.

Breiman, L. (2001). Random forests. *Machine learning*, 45(1): 5–32.

Candanedo, L. M., Feldheim, V. and Deramaix, D. (2017). Data driven prediction models of energy use of appliances in a low-energy house. *Energy and Buildings*, 140: 81–97.

Castro, J. L., Mantas, C. J. and Benítez, J. M. (2000). Neural networks with a continuous squashing function in the output are universal approximators. *Neural Networks*, 13(6): 561–563.

Chang, C. C. and Lin, C. J. (2001). Training v-support vector classifiers: theory and algorithms. *Neural computation*, 13(9): 2119–2147.

Chang, C.-C. and Lin, C.-J. (2011). LIBSVM: A library for support vector machines. *ACM Trans. Intell. Syst. Technol.*, 2(3): Article 27. doi:10.1145/1961189.1961199.

Ekonomou, L. (2010). Greek long-term energy consumption prediction using artificial neural networks. *Energy*, 35(2): 512–517.

Honik, K. (1991). Approximation capabilities of multilayer feedforward network. *Neural Networks*, 4(2): 251–257.

Ilonen, J., Kamarainen, J. K. and Lampinen, J. (2003). Differential evolution training algorithm for feed-forward neural networks. *Neural Processing Letters*, 17(1): 93–105.

Koopialipoor, M. Armaghani, D. J. Hedayat, A. Marto, A. Gordan, B. (2019). Applying various hybrid intelligent systems to evaluate and predict slope stability under static and dynamic conditions. *Soft Computing*, 23: 5913–5929.

Kouziokas, G. N. (2019). Long Short-Term Memory (LSTM) deep neural networks in energy appliances prediction. pp. 1–5. *In: 2019 Panhellenic Conference on Electronics & Telecommunications (PACET)*. IEEE. doi: https://doi.org/10.1109/PACET48583.2019.8956252.

Mirjalili, S., Hashim, S. Z. M. and Sardroudi, H. M. (2012). Training feedforward neural networks using hybrid particle swarm optimization and gravitational search algorithm. *Applied Mathematics and Computation*, 218(22): 11125–11137.

Moayedi, H. Moatamediyan, A. Nguyen, H., Bui, X.-N., Bui, D. T. and Rashid, A. S. A. (2019). Prediction of ultimate bearing capacity through various novel evolutionary and neural network models. *Engineering with Computers*, 1–17.

Moldovan, D. and Slowik, A. (2021). Energy consumption prediction of appliances using machine learning and multi-objective binary grey wolf optimization for feature selection. *Applied Soft Computing*, 111: 107745.

Nourbakhsh, A., Safikhani, H. and Derakhshan, S. (2011). The comparison of multi-objective particle swarm optimization and NSGA II algorithm: applications in centrifugal pumps. *EnOp*, 43: 1095–1113.

Panchal, G., Ganatra, A., Kosta, Y. P. and Panchal, D. (2011). Behaviour analysis of multilayer perceptrons with multiple hidden neurons and hidden layers. *International Journal of Computer Theory and Engineering*, 3(2): 332–337.

Socha, K. and Blum, C. (2007). An ant colony optimization algorithm for continuous optimization: application to feed-forward neural network training. *Neural Computing and Applications*, 16(3): 235–247.

Tsutsui, S., Yamamura, M. and Higuchi, T. (1999). Multi-parent recombination with simplex crossover in real coded genetic algorithms. pp. 657–664. In *Proceedings of the 1st Annual Conference on Genetic and Evolutionary Computation-Volume 1*.

Vapnik, V. N. (1995). The nature of statistical learning. *Theory*. Springer-Verlag New York.

Wang, H., Cui, Z., Chen, Y., Avidan, M., Abdallah, A. B. and Kronzer, A. (2018). Predicting hospital readmission via cost-sensitive deep learning. *IEEE/ACM Transactions on Computational Biology and Bioinformatics*, 15(6): 1968–1978.

Yam, J. Y. and Chow, T. W. (2001). Feedforward networks training speed enhancement by optimal initialization of the synaptic coefficients. *IEEE Transactions on Neural Networks*, 12(2): 430–434.

Zhang, C. and Shao, H. (2000). An ANN's evolved by a new evolutionary system and its application. *Proceedings of the 39th IEEE Conference on Decision and Control (Cat. No. 00CH37187)*, 4: 3562–3563.

CHAPTER 9
Swarm and Evolutionary Intelligence in Deep Learning

◇◇

This chapter is devoted to deep learning optimization applications. A theoretical background discussion of deep unidirectional, bi-directional long short-term memory (LSTM and Bi-LSTM) and Convolutional Neural Networks (CNN) is presented and also topology and weight swarm and evolutionary optimization methods applied on those deep learning methods. The experimental part of this chapter is involved with swarm intelligence and evolutionary computation examples implemented on deep learning models including an illustrative example of covid19 diagnosis from chest x-ray images.

Deep LSTM and Bi-LSTM networks

In the recent past deep unidirectional Long Short-Term Memory (LSTM) networks and Bi-directional LSTM (Bi-LSTM) have been applied in several scientific problems as forecasting networks. The LSTM and Bi-LSTM networks have a topology similar to the Recurrent Neural Networks (RNNs) but the main advantage of LSTM and Bi-LSTM networks over the typical RNNs is that they overcome the vanishing gradients problem which is a very important drawback of the normal RNNs (Cortez et al., 2018). Several researchers have investigated the application of LSTM and Bi-LSTM models in many scientific fields (Behera et al., 2021; Chadha et al., 2020; Kouziokas, 2019).

The LSTM network consists of four major kinds of gates: input gate, forget gate, cell candidate and output gate. The equations which represent the estimation of every component at time step t are as follows.

The equation for the input gate is expressed by the mathematical formula:

$$i_t = \sigma_g(W_i\,x_t + R_i\,h_t - 1 + b_i) \tag{9.1}$$

The equation for the forget gate is expressed by the mathematical formula:

$$f_t = \sigma_g(W_f\,x_t + R_f\,h_t - 1 + b_f) \tag{9.2}$$

The equation for the cell candidate is expressed by the mathematical formula:

$$g_t = \sigma_c(W_g x_t + R_g h_t - 1 + b_g) \tag{9.3}$$

The equation for the output gate is expressed by the mathematical formula:

$$o_t = \sigma_g(W_o\,x_t + R_o\,h_t - 1 + b_o) \tag{9.4}$$

where,

- i, f, g, o represent the input gate, the forget gate, the cell candidate, and the output gate, correspondingly,
- W represents the input weights
- R represents the recurrent weights
- b represents the bias
- h_t the hidden state at time step t
- σ_c the state activation function.

The cell state in time t is expressed by the mathematical formula:

$$c_t = f_t \odot c_{t-1} + i_t \odot g_t \tag{9.5}$$

where: $\odot$ represents the Hadamard product.

The hidden state in a given time t can be expressed by the mathematical formula:

$$c_t = f_t \odot c_{t-1} + i_t \odot g_t \tag{9.6}$$

In the bidirectional LSTM (Bi-LSTM) topology, the network has two directions: forwards and backwards, and it's capable of using data from both sides unlike standard unidirectional LSTM (Schuster and Paliwal, 1997). This helps Bi-LSTM networks to achieve improved forecasting results according to the literature (Shahid et al., 2020; Zhang et al., 2020).

Deep CNN (Convolutional Neural Networks)

CNN is a frequently-used deep learning method in the literature in tasks which have to do with images. Convolutional neural networks are mostly used in image classification and computer vision. Before CNNs, difficult feature extraction methods were utilized to classify objects and images. Convolutional neural networks have improved performance when the inputs are: images or audio signals compared to the other types of neural networks. CNNs have several types of layers, such as: convolutional layers, pooling layers, activation layers, dropout layers, fully-connected layers with several hyperparameters such as number and size of filters, types of pooling (Max pooling or average pooling). The convolutional and pooling layers usually utilize ReLu activation functions, fully-connected layers usually implement a SoftMax activation function in order to classify the inputs (LeCun et al., 2010). An example of the convolutional layer operation is illustrated in Figure 9.1.

The role of the pooling layer is to reduce the complexity of the CNN by down sampling feature maps. The maximum and average poolings are most utilized. A simple example of average pooling is illustrated in Figure 9.2.

A pooled feature map is then flattened into a column as shown in Figure 9.3 in order to be used as a single input vector for the artificial neural network. The role of the neural network in the fully connected layer is to process the input features and predict the result in the output, so that the convolutional network can finally classify the input images.

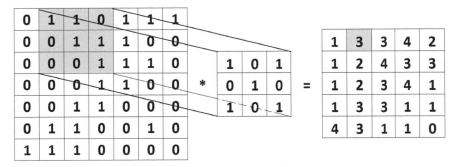

Figure 9.1. Example of convolution function.

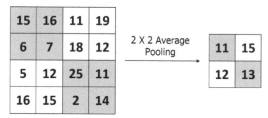

Figure 9.2. Example of convolution function.

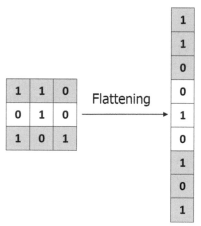

Figure 9.3. Example of a simple flattening operation.

There are several pre-trained models for several kinds of problems that can be used in transfer learning, such as ResNet18 (18 layers), ResNet50 (50 layers), AlexNet, GoogleNet and Darknet19 (LeCun et al., 2015).

CNN and LSTM optimization

Topology optimization

Swarm intelligence and evolutionary strategies can be also implemented to optimize the topology of deep learning models regarding the number of hidden neurons in the hidden layer, the number of filters in the convolutional layers, the pooling type and others.

The objective function should be set for the swarm intelligence or evolutionary method in order to optimize the deep learning model. The most commonly used objective function according to the literature is an error metric. For regression problems the mean squared error or the mean absolute error can be utilized and for classification problems error metrics such as accuracy, sensitivity and F-score are used.

According to the literature, swarm intelligence methods have a better performance in deep learning optimization than other optimization algorithms (Kim and Cho, 2019; Ren et al., 2021; Wang et al., 2020). The deep learning methods can be also optimized in other ways, such as optimizing the training option parameters: Maximum epochs, initial learning rate, optimization algorithm, l_2 regularization value, minibatch size, momentum and others.

Weight optimization

The weight optimization of deep learning methods can be also implemented by using swarm and evolutionary optimization methods (Ilonen et al., 2003; Socha and Blum, 2007). Similarly with topology optimization, a suitable fitness function for the swarm intelligence of the evolutionary method should be set in order to train the deep learning model such as: the mean squared error (MSE) or root mean squared error (RMSE) or mean absolute error (MAE) in regression problems and accuracy, precision, recall, F-score in classification problems.

In the deep learning model weight optimization, the computational complexity is increased. Generally, in fine tuning using transfer learning models such as ResNet18, DarkNet19 the optimization method is applied on the last trainable layer (Koppu et al., 2020). Each particle position represents a solution to the optimization problem of the objective function and the deep learning weight vector is optimized by the swarm particles in order to get the best solution for the fitness function.

Experiments

Bi-LSTM optimization

Dataset

The Japanese Vowels Data Set was utilized for the experiments. The dataset was acquired by the machine learning repository[2] (Kudo et al., 1999). It consists of 640 instances with 12 attributes. The database instances were preprocessed for possible inconsistencies and null values. Then they were

---

[2] https://archive.ics.uci.edu/ml/datasets/Japanese+Vowels

normalized by using min-max normalization and separated into three parts: training, validation, and test set.

Objective function

The percentage error was utilized as the objective function. The accuracy measures the number of correct predictions divided by the total number of the predictions. The percentage error is expressed by the equation:

$$\text{Percentage error} = 1 - \text{Accuracy} = 1 - \frac{Correctly\ predicted\ intances}{Total\ number\ of\ intances} \quad (9.7)$$

Experimental setup

Two methods were utilized as optimization techniques in order to optimize the deep Bidirectional LSTM: adaptive inertia weight particle swarm optimization (Adaptive w-PSO) and genetic algorithm (GA).

Genetic algorithm parameters

Parameter	Value
Population size:	50
Number of generations:	48
Selection:	roulette wheel
Crossover type:	single point
Mutation:	gaussian
Probability of crossover:	0.6
Probability of mutation:	0.033

Adaptive w-PSO parameters

Parameter	Value
Population size:	50
Swarm topology:	Fully connected
Swarm particle initialization	random uniform distribution
Number of iterations:	48

Adaptive inertia weight range: [0.2, 1.1]

Initial inertia weight   max(Inertia Range) – 1.1

Coefficient parameter of personal learning: $c_1 = 2.0$

Coefficient parameter of global learning: $c_2 = 2.0$

Bidirectional LSTM training parameters

Parameter	Value
Minibatch size:	32
epochs:	30
momentum:	0.9
optimizer	adam
Gradient Threshold	1
layers	sequence input layer
	bilstm layer
	fully connected layer
	softmax layer
	classification layer
input feature size	12
number of classes	9

The hyperparameters which were selected to be optimized are: hidden layer neurons, initial learning Rate, L2 regularization value. In the next table the hyperparameter lower and upper bounds in the search space are illustrated.

Table 9.1. Hyperparameter lower and upper bounds in a continuous search space.

Hyperparameter	Lower bound	Upper bound
Hidden layer neurons	100	250
Initial Learning Rate	0.01	0.09
L2 Regularization	0.0001	0.0009

The results of the hyperparameter values and percentage error during adaptive w-PSO-Bi-LSTM optimization are presented in the next table.

Table 9.2. Hyperparameter values and percentage error during adaptive w-PSO-BiLSTM optimization.

Iteration	Hidden Layer Size	Initial Learning Rate	L2 Regularization	Percentage Errors	Accuracy	Best Pe Error So Far	Best Accuracy So Far
1	189	0.0823	0.00064155	0.0811	0.9189	0.0811	0.9189
2	225	0.0458	0.00079433	0.1027	0.8973	0.0811	0.9189
3	106	0.0178	0.00073525	0.0405	0.9595	0.0405	0.9595
4	175	0.0882	0.00029546	0.1432	0.8568	0.0405	0.9595
5	202	0.0796	0.00060226	0.0649	0.9351	0.0405	0.9595
6	138	0.049	0.00012304	0.0757	0.9243	0.0405	0.9595
7	221	0.036	0.00048475	0.0784	0.9216	0.0405	0.9595
8	153	0.0862	0.00049437	0.1216	0.8784	0.0405	0.9595
9	128	0.0477	0.00029818	0.0595	0.9405	0.0405	0.9595
10	233	0.0876	0.00020224	0.2162	0.7838	0.0405	0.9595
11	198	0.0694	0.00088528	0.0919	0.9081	0.0405	0.9595
12	219	0.0201	0.00048803	0.0405	0.9595	0.0405	0.9595
13	112	0.0189	0.00019942	0.0351	0.9649	0.0351	0.9649
14	203	0.0632	0.00043017	0.1432	0.8568	0.0351	0.9649
15	122	0.0156	0.00089184	0.1135	0.8865	0.0351	0.9649
16	183	0.0508	0.00087723	0.0892	0.9108	0.0351	0.9649
17	145	0.0342	0.00080962	0.0514	0.9486	0.0351	0.9649
18	185	0.0761	0.00016918	0.1703	0.8297	0.0351	0.9649
19	124	0.0895	0.00026703	0.0649	0.9351	0.0351	0.9649
20	104	0.0605	0.00084476	0.0946	0.9054	0.0351	0.9649
21	125	0.0169	0.00074574	0.0595	0.9405	0.0351	0.9649
22	170	0.0239	0.00037608	0.0676	0.9324	0.0351	0.9649
23	120	0.0386	0.00027356	0.0514	0.9486	0.0351	0.9649
24	236	0.0653	0.00081053	0.0865	0.9135	0.0351	0.9649
25	207	0.0531	0.00060504	0.1459	0.8541	0.0351	0.9649
26	223	0.0701	0.00030383	0.0973	0.9027	0.0351	0.9649
27	217	0.0284	0.00062058	0.0568	0.9432	0.0351	0.9649
28	119	0.0662	0.00020829	0.0622	0.9378	0.0351	0.9649
29	122	0.0818	0.00017933	0.0405	0.9595	0.0351	0.9649
30	134	0.0411	0.00080546	0.0541	0.9459	0.0351	0.9649
31	191	0.0151	0.00051611	0.0351	0.9649	0.0351	0.9649
32	150	0.0339	0.00039142	0.0784	0.9216	0.0351	0.9649
33	135	0.0537	0.00024335	0.0973	0.9027	0.0351	0.9649
34	113	0.0423	0.00013362	0.0649	0.9351	0.0351	0.9649
35	226	0.0725	0.00016819	0.2568	0.7432	0.0351	0.9649
36	219	0.0634	0.00031599	0.1486	0.8514	0.0351	0.9649
37	113	0.0113	0.00014789	0.0405	0.9595	0.0351	0.9649

Table 9.2 contd. ...

...Table 9.2 contd.

Iteration	Hidden Layer Size	Initial Learning Rate	L2 Regularization	Percentage Errors	Accuracy	Best Pe Error So Far	Best Accuracy So Far
38	153	0.0602	0.00079502	0.0595	0.9405	0.0351	0.9649
39	202	0.0147	0.00042565	0.0378	0.9622	0.0351	0.9649
40	243	0.0524	0.00052815	0.0351	0.9649	0.0351	0.9649
41	104	0.0678	0.00085988	0.0703	0.9297	0.0351	0.9649
42	193	0.0466	0.00066194	0.1784	0.8216	0.0351	0.9649
43	215	0.034	0.00019361	0.0757	0.9243	0.0351	0.9649
44	170	0.03	0.00024958	0.0919	0.9081	0.0351	0.9649
45	146	0.0199	0.00070955	0.0324	0.9676	0.0324	0.9676
46	210	0.0348	0.00088012	0.0297	0.9703	0.0297	0.9703
47	180	0.0209	0.00070588	0.0703	0.9297	0.0297	0.9703
48	123	0.0627	0.0002885	0.0676	0.9324	0.0297	0.9703

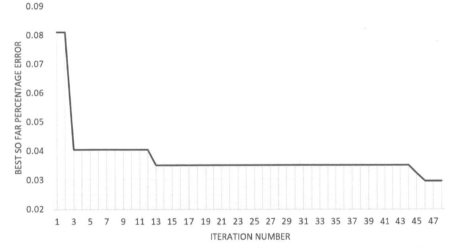

Figure 9.4. Best so far classification error plot during the iterations by using PSO to optimize Bi-LSTM hyperparameters.

The results of the hyperparameter values and percentage error during adaptive w-GA-Bi-LSTM optimization are presented in the next table.

As we can observe from the results the PSO optimization of Bi-LSTM network is much faster than the GA-Bi-LSTM model. Within 3 iterations PSO achieves a percentage error of about 0.04 much smaller than the GA

Table 9.3. Hyperparameter values and classification error during GA-BiLSTM optimization.

Iteration	Hidden Layer Size	Initial Learning Rate	L2 Regularization	Percent Errors	Accuracy	Best % error so far	Best Accuracy so far
1	100	0.01	0.0009	0.07027	0.92973	0.07027	0.92973
2	212	0.053998	0.00084847	0.062162	0.93784	0.062162	0.937838
3	187	0.019514	0.00055323	0.048649	0.95135	0.048649	0.951351
4	190	0.071739	0.00087868	0.062162	0.93784	0.048649	0.951351
5	249	0.059428	0.00030971	0.33784	0.66216	0.048649	0.951351
6	214	0.080309	0.00013372	0.072973	0.92703	0.048649	0.951351
7	129	0.072257	0.00056525	0.078378	0.92162	0.048649	0.951351
8	142	0.040555	0.00053645	0.059459	0.94054	0.048649	0.951351
9	140	0.035934	0.00075752	0.054054	0.94595	0.048649	0.951351
10	237	0.075581	0.00077861	0.13514	0.86486	0.048649	0.951351
11	210	0.028916	0.00057152	0.064865	0.93514	0.048649	0.951351
12	221	0.050018	0.00010082	0.051351	0.94865	0.048649	0.951351
13	144	0.087712	0.00070494	0.13784	0.86216	0.048649	0.951351
14	122	0.028447	0.00075638	0.054054	0.94595	0.048649	0.951351
15	230	0.043063	0.00080923	0.091892	0.90811	0.048649	0.951351
16	225	0.088903	0.00060577	0.045946	0.95405	0.045946	0.954054
17	161	0.064449	0.00063802	0.032432	0.96757	0.032432	0.967568
18	225	0.022482	0.00076005	0.043243	0.95676	0.032432	0.967568
19	122	0.060157	0.00011498	0.051351	0.94865	0.032432	0.967568
20	220	0.052159	0.00087828	0.11351	0.88649	0.032432	0.967568
21	247	0.052241	0.00072428	0.07027	0.92973	0.032432	0.967568
22	106	0.020415	0.00038129	0.027027	0.97297	0.027027	0.972973
23	146	0.04792	0.00027676	0.086486	0.91351	0.027027	0.972973
24	216	0.067964	0.00057967	0.1027	0.8973	0.027027	0.972973
25	247	0.020178	0.00033221	0.1000	0.9	0.027027	0.972973
26	129	0.01191	0.00026627	0.043243	0.95676	0.027027	0.972973
27	182	0.044048	0.0001073	0.032432	0.96757	0.027027	0.972973
28	237	0.067764	0.00055862	0.11892	0.88108	0.027027	0.972973
29	124	0.039579	0.00020074	0.083784	0.91622	0.027027	0.972973
30	193	0.061625	0.00072365	0.056757	0.94324	0.027027	0.972973
31	139	0.054479	0.00017531	0.07027	0.92973	0.027027	0.972973
32	243	0.02894	0.00049466	0.067568	0.93243	0.027027	0.972973
33	132	0.03971	0.00044401	0.045946	0.95405	0.027027	0.972973
34	181	0.027902	0.00054813	0.048649	0.95135	0.027027	0.972973
35	162	0.081133	0.00082185	0.091892	0.90811	0.027027	0.972973
36	102	0.07662	0.00036034	0.14865	0.85135	0.027027	0.972973

Table 9.3 contd. ...

...*Table 9.3 contd.*

| 37 | 237 | 0.054978 | 0.00068724 | 0.035135 | 0.96486 | 0.027027 | 0.972973 |
Iteration	Hidden Layer Size	Initial Learning Rate	L2 Regularization	Percent Errors	Accuracy	Best % error so far	Best Accuracy so far
38	240	0.043808	0.00044434	0.054054	0.94595	0.027027	0.972973
39	158	0.080646	0.00086454	0.072973	0.92703	0.027027	0.972973
40	224	0.076497	0.00036992	0.15405	0.84595	0.027027	0.972973
41	108	0.037876	0.00020802	0.037838	0.96216	0.027027	0.972973
42	208	0.051865	0.00031076	0.089189	0.91081	0.027027	0.972973
43	116	0.059506	0.00015485	0.062162	0.93784	0.027027	0.972973
44	249	0.061759	0.00042695	0.11622	0.88378	0.027027	0.972973
45	241	0.075807	0.0004411	0.15676	0.84324	0.027027	0.972973
46	122	0.079736	0.00013518	0.067568	0.93243	0.027027	0.972973
47	146	0.086798	0.00030496	0.081081	0.91892	0.027027	0.972973
48	154	0.03394	0.00024518	0.043243	0.95676	0.027027	0.972973

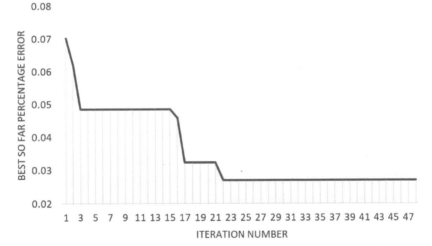

Figure 9.5. Best so far classification error plot during the iterations by using GA-Bi-LSTM model.

which achieves about 0.05 in the same iteration number, although, the results are similar in 48 iterations.

CNN optimization

CNN – PSO model

In the experiments, the CNN model consists of a convolutional layer, ReLU activation function and a pooling layer, a second convolutional layer, and ReLU activation and a second pooling layer and a dropout layer. The fully connected layer consists of an input layer, a hidden layer, a ReLU activation function and an output layer with the softmax function as the activation function. The kernel size of pooling layers was 2 × 2. All the PSO parameters are randomly initialized. The initial learning rate was 0.001 with the Adam optimizer. The hyperparameters to be optimized by the adaptive PSO where: Number of filters in every convolutional layer (1–100), size of trainable filters in every convolutional layer (3–10), type of pooling in every pooling layer (average pooling and max-pooling), probability of dropout in every convolutional layer (0.1–0.9), batch size (16–128), maximum epochs (30–50), and number of hidden neurons in the hidden layer of the fully connected layer (50–400).

Evaluation metrics

This section lists the evaluation metrics used to measure the classification performance of the experimental models.

Precision

The precision metric evaluates the prediction accuracy and it is expressed by the following equation:

$$precision = \frac{TP}{TP + FP} \tag{9.8}$$

where,

- TP represents the true positive cases
- FP for the false positive cases.

Recall or sensitivity

The recall metric, also referred to as sensitivity, evaluates the results completeness of classification and it is calculated by the following equation:

$$Recall = \frac{TP}{TP + FN} \tag{9.9}$$

where,

- TP represents the true positive cases
- FN represents the false negative cases.

F-measure

The F-Measure metric evaluates the prediction performance of a classification algorithm, by combining precision and recall measures and is calculated by the following equation:

$$F\ measure = 2 * \frac{precision * recall}{precision + recall} \quad (9.10)$$

By combining the previous equations, the F-Measure can be expressed by the following equation:

$$F\ measure = 2 * \frac{TP}{2 * TP + FP + FN} \quad (9.11)$$

Matthews correlation coefficient (MCC)

The Matthews Correlation Coefficient is considered as a balanced classification measure and it is estimated by the following mathematical equation:

$$MCC = \frac{TP * TN - FP * FN}{[(TP + FP) * (FN + TN) * (FP + TN) * (TP + FN)]^{\frac{1}{2}}} \quad (9.12)$$

where TN represents the true negative cases.

Receiver operating characteristic (ROC)

A receiver operating characteristic curve, is a graphical plot that illustrates the ability of a classifier. The ROC curve is designed by plotting the true positive rate against the false positive rate instances for various threshold configurations.

Precision-recall curve (PRC)

A precision-recall curve is a plot designed between the precision metric (y-axis) and the recall metric (x-axis) for various threshold configurations, similar to the ROC curve.

Covid-19 chest X-ray dataset

The chest x-ray images of the ieee8023 dataset were utilized (Cohen et al., 2020). The chest x-ray images were preprocessed by using dynamic histogram equalization for image contrast enhancement (Abdullah-Al-

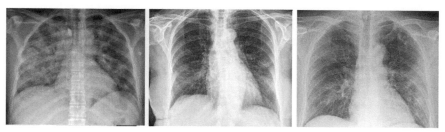

Figure 9.6. Some x-ray images of patients with covid-19 from the ieee8023 dataset (Cohen et al., 2020).

Wadud et al., 2007). The dataset was split to training and test sets. The test set has 66 images from every class: normal, covid19 and other diseases.

Experimental results

CNN without PSO

In the CNN model the overall correctly classified instances in the test set were 167 (84.34%) and the incorrectly classified instances were 31 (15.65%). The classification metrics for the CNN model without PSO optimization are presented in the Table 9.4. The confusion matrix is illustrated in Table 9.5.

Table 9.4. Results for the CNN without PSO optimization.

Class	TP Rate	FP Rate	Precision	Recall	F-Measure	MCC	ROC Area	PRC Area
COVID-19	0,788	0,038	0,912	0,788	0,846	0,781	0,922	0,917
Normal	0,909	0,068	0,870	0,909	0,889	0,832	0,954	0,819
Other diseases	0,833	0,129	0,764	0,833	0,797	0,690	0,922	0,859
Weighted Avg.	0,843	0,078	0,849	0,843	0,844	0,768	0,933	0,865

Table 9.5. Confusion matrix for the CNN without PSO optimization.

		PREDICTED		
		COVID-19	Normal	Other diseases
ACTUAL	COVID-19	52	3	11
	Normal	0	60	6
	Other diseases	5	6	55

CNN optimized with PSO

Adaptive PSO parameters

The adaptive inertia weight PSO parameters for the experiments were the following:

Parameter	Value
Population size:	50
Swarm topology:	Fully connected
Swarm particle initialization	random uniform distribution
Number of iterations:	50
Adaptive inertia weight range:	[0.2, 1.1]
Initial inertia weight	max(Inertia Range) = 1.1
Coefficient parameter of personal learning:	$c_1 = 2.0$
Coefficient parameter of global learning:	$c_2 = 2.0$

In the CNN model optimized with PSO, the overall correctly classified instances in the test set were 183 (92.42%) and the incorrectly classified instances were 15 (7.58%). The classification metrics for the CNN model without PSO optimization are presented in the Table 9.6. The confusion matrix is illustrated in Table 9.7.

Table 9.6. Results for the CNN - PSO model.

Class	TP Rate	FP Rate	Precision	Recall	F-Measure	MCC	ROC Area	PRC Area
COVID-19	0,879	0,015	0,967	0,879	0,921	0,886	0,932	0,890
Normal	0,970	0,045	0,914	0,970	0,941	0,911	0,962	0,897
Other diseases	0,924	0,053	0,897	0,924	0,910	0,865	0,936	0,854
Weighted Avg.	0,924	0,038	0,926	0,924	0,924	0,887	0,943	0,880

Table 9.7. Confusion matrix for the CNN - PSO model.

		PREDICTED		
		COVID-19	Normal	Other diseases
ACTUAL	COVID-19	58	3	5
	Normal	0	64	2
	Other diseases	2	3	61

As we can observe the PSO optimized CNN network produced much better results than the CNN model without any optimization.

This is a simple experiment of the application of swarm intelligence on deep learning To achieve more accurate results other optimized deep learning models can be used. For example, transfer learning can be utilized to train the last layers of pre-trained models like ResNet18, ResNet50, AlexNet, GoogleNet and Darknet19 (LeCun et al., 2015) and swarm intelligence models can be used to optimize either the layer weights or the hyperparameters of the deep learning models such as: hidden layer neurons of the fully connected layer, the initial learning rate, the L2 regularization factor, the number and size of filters in the convolutional layers, the maximum epochs, the batch size or other parameters. Furthermore, other techniques can be applied to improve the results by changing: the network weight initialization method, the normalization method, the validation method (E.g., by using 10-Fold Cross-Validation), the training algorithm, the feature selection method or use a feature extraction technique. In addition data handling methods can be used to create a balanced dataset from an imbalanced dataset like the Synthetic Minority Oversampling Technique (SMOTE) (Chawla et al., 2002). Also, data augmentation methods can be implemented to increase the dataset instances and improve the generalization and the accuracy of the deep learning models.

References

Abdullah-Al-Wadud, M., Kabir, M. H., Dewan, M. A. A. and Chae, O. (2007). A dynamic histogram equalization for image contrast enhancement. *IEEE Transactions on Consumer Electronics*, 53(2): 593–600. doi:10.1109/TCE.2007.381734.

Behera, S., Misra, R. and Sillitti, A. (2021). Multiscale deep bidirectional gated recurrent neural networks based prognostic method for complex non-linear degradation systems. *Information Sciences*, 554: 120–144.

Chadha, G. S., Panambilly, A., Schwung, A. and Ding, S. X. (2020). Bidirectional deep recurrent neural networks for process fault classification. *ISA Transactions*, 106: 330–342.

Chawla, N. V., Bowyer, K. W., Hall, L. O. and Kegelmeyer, W. P. (2002). SMOTE: synthetic minority over-sampling technique. *Journal of Artificial Intelligence Research*, 16: 321–357.

Cohen, J. P., Morrison, P. and Dao, L. (2020). COVID-19 image data collection. *arXiv preprint arXiv:2003.11597*. https://github.com/ieee8023/covid-chestxray-dataset.

Cortez, B., Carrera, B., Kim, Y. J. and Jung, J. Y. (2018). An architecture for emergency event prediction using LSTM recurrent neural networks. *Expert Systems with Applications*, 97: 315–324.

Ilonen, J., Kamarainen, J. K. and Lampinen, J. (2003). Differential evolution training algorithm for feed-forward neural networks. *Neural Processing Letters*, 17(1): 93–105.

Kim, T. Y. and Cho, S. B. (2019). Particle swarm optimization-based CNN-LSTM networks for forecasting energy consumption. pp. 1510–1516. *In: 2019 IEEE Congress on Evolutionary Computation (CEC)*. IEEE.

Koppu, S., Maddikunta, P. K. R. and Srivastava, G. (2020). Deep learning disease prediction model for use with intelligent robots. *Computers & Electrical Engineering*, 87: 106765.

Kouziokas, G. N. (2021). Deep bidirectional and unidirectional LSTM neural networks in traffic flow forecasting from environmental factors. *Advances in Mobility-as-a-Service Systems*, 1278: 171–180. doi:https://doi.org/10.1007/978-3-030-61075-3_17.

Kudo, M., Toyama, J. and Shimbo, M. (1999). Multidimensional curve classification using passing-through regions. *Pattern Recognition Letters*, 20(11-13): 1103–1111.

LeCun, Y., Kavukcuoglu, K. and Farabet, C. (2010). Convolutional networks and applications in vision. pp. 253–256. *In: Proceedings of 2010 IEEE International Symposium on Circuits and Systems*. IEEE.

LeCun, Y., Bengio, Y. and Hinton, G. (2015). Deep learning. *Nature*, 521(7553): 436–444.

Ren, X., Liu, S., Yu, X. and Dong, X. (2021). A method for state-of-charge estimation of lithium-ion batteries based on PSO-LSTM. *Energy*, 234: 121236.

Schuster, M. and Paliwal, K. K. (1997). Bidirectional recurrent neural networks. *IEEE Transactions on Signal Processing*, 45(11): 2673–2681.

Shahid, F., Zameer, A. and Muneeb, M. (2020). Predictions for COVID-19 with deep learning models of LSTM, GRU and Bi-LSTM. *Chaos, Solitons & Fractals*, 140: 110212.

Socha, K. and Blum, C. (2007). An ant colony optimization algorithm for continuous optimization: application to feed-forward neural network training. *Neural Computing and Applications*, 16(3): 235–247.

Wang, P., Zhao, J., Gao, Y., Sotelo, M. A. and Li, Z. (2020). Lane work-schedule of toll station based on queuing theory and PSO-LSTM model. *IEEE Access*, 8: 84434–84443.

Zhang, B., Zhang, H., Zhao, G. and Lian, J. (2020). Constructing a PM2. 5 concentration prediction model by combining auto-encoder with Bi-LSTM neural networks. *Environmental Modelling & Software*, 124: 104600.

Index